Synthesis Lectures on Power Electronics

This series publishes short books on topics related to power electronics, ancillary components, packaging and integration, electric machines and their drive systems, as well as related subjects such as EMI and power quality. Each Lecture develops a particular topic with the requisite introductory material and progresses to more advanced subject matter such that a comprehensive body of knowledge is encompassed. Simulation and modeling techniques and examples are included where applicable.

Nie Hou

High-Robust Control Schemes for Dual-Active-Bridge-Based DC–DC Converter Systems in Renewable Energy Applications

 Springer

Nie Hou
Department of Engineering
University of Alberta
Edmonton, AB, Canada

ISSN 1931-9525 ISSN 1931-9533 (electronic)
Synthesis Lectures on Power Electronics
ISBN 978-3-031-72962-1 ISBN 978-3-031-72963-8 (eBook)
https://doi.org/10.1007/978-3-031-72963-8

This Springer imprint is published by the registered company Springer Nature Switzerland AG
The registered company address is: Gewerbestrasse 11, 6330 Cham, Switzerland

If disposing of this product, please recycle the paper.

Supervisor's Foreword

With the rapid development of renewable energies such as wind and solar energy, the DC microgrid has gained significant attention as an efficient method for integrating distributed renewable resources with less power conversion stages and without issues such as harmonics and synchronization. Within a DC microgrid, a DC–DC converter is the primary component for realizing power transmission and maintaining DC-link voltages, with the dual-active-bridge (DAB), DC–DC converter gained the most attention in the past 10 years due to merits such as high efficiency, bidirectional operation, and excellent controllability.

Although the DAB converter has been extensively investigated, the research on other DC–DC converters, such as full bridge DC–DC converter and three-phase DAB converter, has been seldom covered, especially in terms of robust control. Generally, these neglected converters transfer power based on the AC inductance, thereby naming them as intermediary inductive DC-link (I^2ACL) DC–DC converters in this thesis. Subsequently, based on the analysis of the dynamic equivalence among these converters, a unified fast-dynamic control scheme is proposed for boosting the dynamic response of the I^2ACL DC–DC converters, ultimately reducing the research gap between the DAB DC–DC converter and I^2AC DC–DC converters.

Furthermore, in DC microgrids, a single DAB DC–DC converter typically fails to meet the requirements for high power and voltage, so the modularized DAB converter systems such as input-parallel output-parallel (IPOP), input-independent output-parallel (IIOP), input-parallel output-series (IPOS), and input-series output-parallel (ISOP) configurations are often employed. However, the research on these modularized converter systems is also limited. Consequently, in this thesis, various systematic and advanced control schemes including soft start-up operation, fast dynamic control, flexible power sharing operation, and hot-swap control are proposed for boosting the robustness of these multiple DAB DC–DC converter systems.

This thesis also expands the scope of applications for DAB-based converter systems to integrating the renewable energy source and the energy storage system. Based partial power processing, a DAB-based integrated PV-battery converter system, is proposed for concurrently realizing the maximum power tracking of PV panels and the stabilization of the total DC-link voltage. Compared with existing PV-battery converter systems, the utilization of solar energy can be boosted significantly. Besides, a high-robustness control strategy is proposed for maintaining the DC-link voltage, even when the solar panel is out of service.

This thesis reviews I^2ACL DC–DC converters with dynamic equivalence as DAB converters and presents high robust controls for the DAB-based converter systems. Regarding any errors and inadequacies in the thesis, I sincerely request readers and colleagues to criticize and correct them.

Edmonton, AB, Canada   Prof. Yunwei Li
January 2024

Preface

With the development of renewable energies, such as wind energy and solar energy, the DC power system becomes a promising candidate to manage and transfer the renewable energy source, which stimulates the study of the DC–DC converters in the past decades. Among various DC–DC converters, the dual-active-bridge (DAB) DC–DC converter is regarded as one of the most promising candidates for the DC power conversion due to merits like isolated, high-efficiency, bidirectional, and ultrafast dynamic characteristics.

Except for the DAB DC–DC converter, there are some other isolated DC–DC converters such as full bridge DC–DC converter, three-phase DAB DC–DC converter, etc. They normally have similar dynamic characteristics as the DAB DC–DC converter featuring intermediary inductive AC-link (I^2ACL) configuration. However, they are rarely investigated in existing literature, especially for better dynamic control performance. To fill such a gap, the dynamic equivalence between the DAB DC–DC converter and other I^2ACL isolated DC–DC converters is revealed with the thorough overview of the existing I^2ACL topologies in this work. Further, a unified fast-dynamic direct-current control scheme is proposed for significantly improving the dynamic performance of these I^2ACL isolated DC–DC converters. With this predetermined analysis, the dynamic control schemes for the DAB-based DC–DC converter systems can be easily extended to other I^2ACL converters with the same configurations.

The single DAB DC–DC converter has been extensively investigated, but its modularized converter systems such as input-parallel output-parallel (IPOP), input-independent output-parallel (IIOP), input-parallel output-series (IPOS), and input-series output-parallel (ISOP) configurations have been seldomly covered in the existing research. Particularly, it is emergent to improve the dynamics, e.g. the input-voltage disturbance, the load-condition change and the power sharing disturbance. In this work, the advanced dynamic controls for these modular DAB DC–DC converter systems are proposed, featuring the flexible power sharing control performances with fast-dynamic responses. Moreover, to realize the reliable operation of these DAB-based systems, the hot swap operations are presented. To ensure the desired power sharing performance, circuit-parameter estimating methods are proposed for these DAB-based converter systems.

This work expands the scope of the application of the DAB-based converter system in partial power processing (PPP). Different from the existing literatures focusing on embedding renewable energy source into the strong AC system, this work proposes a PPP converter system, which can realize the independent control of the renewable energy source and the stabilization of the total DC bus. Combining with DAB module, the DAB-based PPP converter system is proposed. Then, as one of the important functions, the stabilization of the total DC bus should be further improved for this DAB-based converter system. In detail, a high-robustness control strategy is proposed to realize the fast-dynamic control, and the operation when one renewable energy source is out of work is also presented. Notably, renewable energy should feature the current output and the limited output-voltage regulation such as PV, fuel cell, and wind turbine with AC-DC conversion. By using the PV as an example, the effectiveness of the novel system is verified with the following results: (1). The maximum power point tracking of the PV panels can be realized by using the existing method. (2). By using the proposed high-robustness control scheme, the total DC-link voltage can maintain at its desired value when the irradiance of PV panels, the voltage of the battery, and the load condition are changed, and even when the PV panel is heavily shaded.

Edmonton, Canada   Nie Hou

Acknowledgements

I would like to express my deep gratitude to all those who have offered valuable inspiration and selfless support during my Ph.D. program.

First and foremost, I would like to thank my Ph.D. supervisor, Prof. Yunwei (Ryan) Li for his continuous support and guidance through these years. His insightful advice and encouragement helped me a lot to overcome different challenges and enrich my research experience. It is always my great honor to join Prof. Li's group and I pretty enjoy this journey.

Secondly, for all the lab mates that have provided their professional suggestions, comments, assistance, please accept my appreciation: Dr. Yuzhuo Li, Dr. Hao Tian, Dr. Ding Li, Dr. Zhongyi Quan, Dr. Fanxiu Fang, Dr. Yue Zhang, Cheng Xue, Mingzhe Wu, Pasan Gunawardena Loku Hettige, Dr. Dulika Nayanasiri, Rouzbeh Reza Ahrabi, Andrew Zhou, Xuesong Wu, etc. I am honored to study and work with all of you, thank you!

Thirdly, I would also like to thank the Alberta Innovates and Natural Sciences and Engineering Research Council of Canada for their financial support during my Ph.D.

Finally, I want to thank my family for their understanding, patience, and love that make this work possible. Especially my wife Mrs. Lisha Liu, I would like to sincerely thank her for her great support and company during the past years, I really enjoy the daily stories we had made and look forward to the future life we planned together.

List of Publications

Chapter 2:

1. **N. Hou**, Y. Li, Z. Quan, Y. W. Li and A. Zhou, "Unified Fast-Dynamic Direct-Current Control Scheme for Intermediary Inductive AC-Link Isolated DC–DC Converters," in *IEEE Open Journal of Power Electronics*, vol. 2, pp. 383–400, 2021.

Chapter 3:

2. **N. Hou** and Y. Li, "The Comprehensive Circuit-Parameter Estimating Strategies for Output-Parallel Dual-Active-Bridge DC–DC Converters with Tunable Power Sharing Control," in *IEEE Transactions on Industrial Electronics*, vol. 67, no. 9, pp. 7583–7594, Sept. 2020.

3. **N. Hou** and Y. Li, "Communication-Free Power Management Strategy for the Multiple DAB-Based Energy Storage System in Islanded DC Microgrid," in *IEEE Transactions on Power Electronics*, vol. 36, no. 4, pp. 4828–4838, April 2021.

Chapter 4:

4. **N. Hou** and Y. W. Li, "A Tunable Power Sharing Control Scheme for the Output-Series DAB DC–DC System with Independent or Common Input Terminals," in *IEEE Transactions on Power Electronics*, vol. 34, no. 10, pp. 9386–9391, Oct. 2019.

5. **N. Hou**, P. Gunawardena, X. Wu, L. Ding, Y. Zhang and Y. W. Li, "An Input-Oriented Power Sharing Control Scheme with Fast-Dynamic Response for ISOP DAB DC–DC Converter," in *IEEE Transactions on Power Electronics*, vol. 37, no. 6, pp. 6501–6510, June 2022.

Chapter 5:

6. **N. Hou**, L. Ding, P. Gunawardena, T. Wang, Y. Zhang and Y. W. Li, "A Partial Power Processing Structure Embedding Renewable Energy Source and Energy Storage Element for Islanded DC Microgrid," in *IEEE Transactions on Power Electronics*, vol. 38, no. 3, pp. 4027–4039, March 2023.

Contents

Abbreviations

CCL	Constant Current Load
CPE	Circuit-Parameter Estimating
CPL	Constant Power Load
DAB	Dual-Active-Bridge
DPP	Differential Power Processing
ESS	Energy Storage System
ESU	Energy Storage Unit
FDDC	Unified Fast-Dynamic Direct-Current
I^2ACL	Intermediary Inductive AC-Link
IIOP	Input-Independent Output-Parallel
IPOP	Input-Parallel Output-Parallel
IPOS	Input-Parallel Output-Series
ISOP	Input-Series Output-Parallel
LVDC	Low-Voltage DC
MPPT	Maximum Power Point Tracking
MVDC	Medium Voltage DC
PI	Proportional-Integral
PPP	Partial Power Processing
PV	Photovoltaic
PV-IP	PV-to-Isolated Port
SPS	Single-Phase-Shift
TPS	Triple-Phase-Shift
VPI	Voltage PI-Based
WT	Wind Turbine

List of Figures

List of Tables

Introduction

Along with the development of renewable energies such as wind energy and solar energy, the dc power distribution system has been extensively investigated for collecting and transferring these energies, which stimulates the research of the dc-dc converters in the past decades. Among various dc-dc converters, the dual-active-bridge (DAB) dc-dc converter is regarded as one of the most promising candidates in the dc power conversion systems because of with symmetric, isolated, high-efficiency, bidirectional and ultrafast dynamic characteristics. The topology of the DAB dc-dc converter can be shown in Fig. 1.1 [1]. This converter has been widely adopted in distributed generating systems [2, 3], automotive applications [4, 5], energy storage systems [6, 7] and power electronic transformer in railway traction applications [8].

Meanwhile, some isolated dc-dc converters such as full bridge dc-dc converter and three-phase DAB dc-dc converter have been studied, which have similar dynamic characteristics as the DAB dc-dc converter featuring intermediary inductive ac-link (I^2ACL) configuration [9, 10]. However, these I^2ACL dc-dc converters do not obtain enough study, especially in terms of the dynamic control. Besides, the dynamic equivalence between the DAB dc-dc converter and other I^2ACL dc-dc converters should also be revealed. Then, with this fundamental analysis, the dynamic controls schemes for the DAB-based dc-dc converter systems can be easily extended to the other I^2ACL dc-dc converters with the same configurations.

Currently, the DAB-based dc-dc converter systems such as input-parallel output-parallel (IPOP), input-independent output-parallel (IIOP) input-parallel output-series (IPOS) and input-series output-parallel (ISOP) configurations are starting to get research on the wind farm, the electric vehicle charge station, the dc power transformer, and the dc microgrid [11–15]. This existing literature is more focused on some basic analysis of

© The Author(s), under exclusive license to Springer Nature Switzerland AG 2025   1
N. Hou, *High-Robust Control Schemes for Dual-Active-Bridge-Based DC–DC Converter Systems in Renewable Energy Applications*, Synthesis Lectures on Power Electronics,
https://doi.org/10.1007/978-3-031-72963-8_1

Fig. 1.1 The topology of the
DAB dc-dc converter

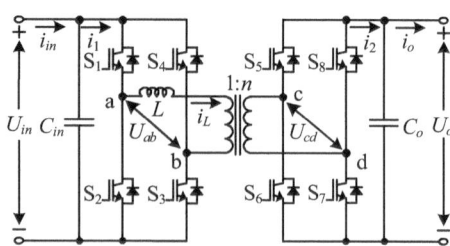

these DAB-based dc-dc converter systems. To deal with the frequent disturbances such as the input-voltage changes and the load changes, the dynamic improvements of these modular DAB dc-dc converter systems need further study. Furthermore, another potentially important application of the DAB-based dc-dc converter system is the partial power processing (PPP) converter system. In the past decades, the PPP converter systems are extensively studied for the strong ac system [16, 17], but the PPP converter system for the islanded dc grid still needs of study. Thus, for the islanded grid system, a DAB-based PPP system is proposed for realizing the independent control of the renewable energy source and the stabilization of the total dc-voltage bus.

In this Chapter, the applications of the I^2ACL dc-dc converters such as the full-bridge dc-dc converter and the DAB dc-dc converter are discussed and analyzed. Besides, the dynamic equivalence in these I^2ACL dc-dc converters is simply introduced in Sect. 1.1. Moreover, the existing studies of the modular DAB dc-dc converter systems including IPOP, IIOP, IPOS and ISOP are introduced and discussed, and the drawbacks of some existing dynamic control schemes are analyzed in Sects. 1.2 and 1.3. In addition, another potentially important application of the DAB-based dc-dc converter system is PPP applications. Different from the existing literatures focusing on the strong ac grid system, this work proposes a DAB-based PPP converter system for the islanded dc microgrid, featuring the independent control of the renewable energy source and the stabilization of the total dc-link voltage in Sect. 1.4. Furthermore, the objectives and the contributions of this work are summarized is Sect. 1.5. Then, the general logic structure of this Chapter can be summarized in Fig. 1.2.

1.1   The I^2ACL Isolated DC-DC Converters

Isolated dc-dc converters have been extensively applied in modern industrial applications such as metro vehicles [18], electric vehicles [4, 19, 20], data center [21], and grid systems [5, 22–24], etc. Real-world projects can be found like the auxiliary power supply of a metro vehicle system shown in Fig. 1.3, where the isolated dc-dc power conversion stage is employed to replace the traditional line-frequency for lower cost, smaller size, and less acoustic noise [18]. Or the utility-service equipment like the back-to-back converter system can be shown in Fig. 1.4, which is embedded with isolated dc-dc converters to

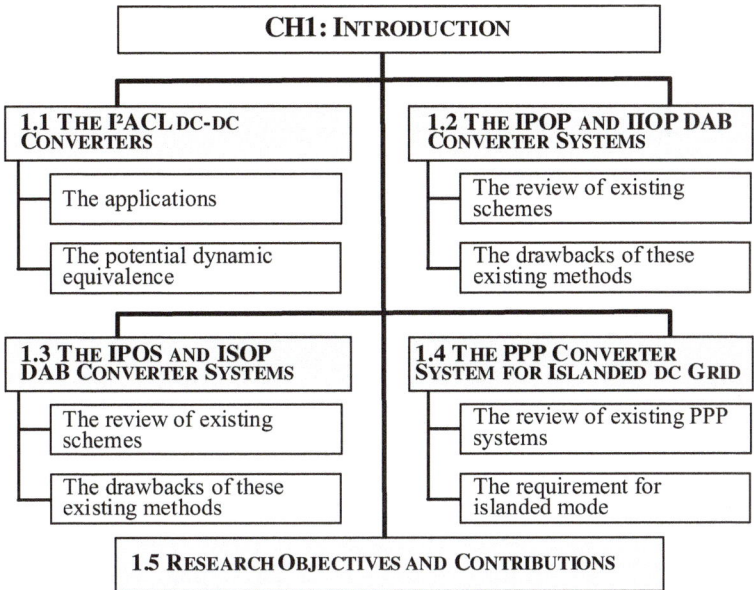

Fig. 1.2 Logic structure of this chapter

solve the power flow balancing problems with asynchronized ac grids [23]. In addition, given the great potentials of dc microgrids over conventional ac configurations, more isolated dc-dc converters can be expected to be employed in the dc microgrid as shown in Fig. 1.5 [24].

In practice, galvanic isolation is required mainly for safety and grounding reasons. The isolated structures offer great flexibility of various grounding techniques on both dc sides, as well as easier parallel or series connections [25, 26]. Therefore, the same elementary

Fig. 1.3 Overall system configuration of 210 kW auxiliary power supply for metro vehicle system

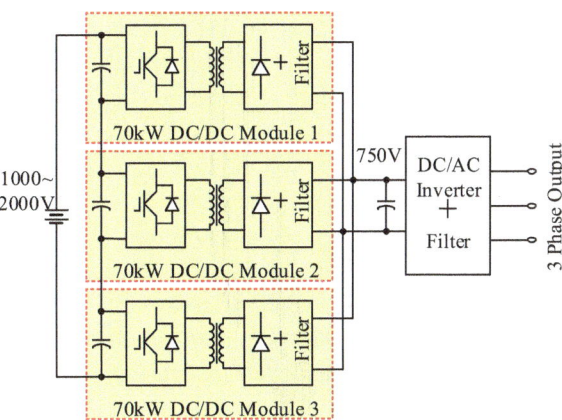

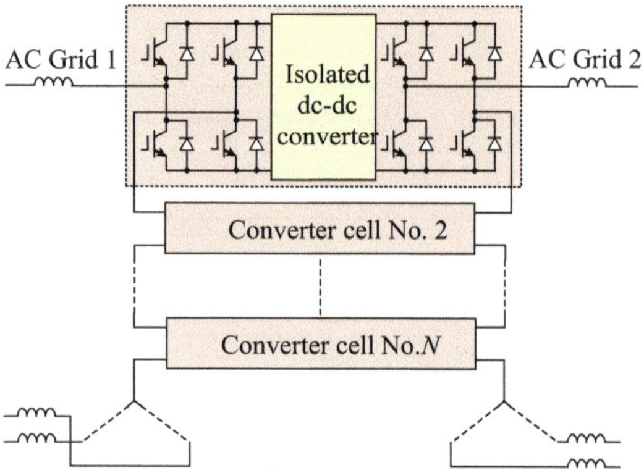

Fig. 1.4 The 6.6 kV back-to-back system in the next-generation medium-voltage power conversion system

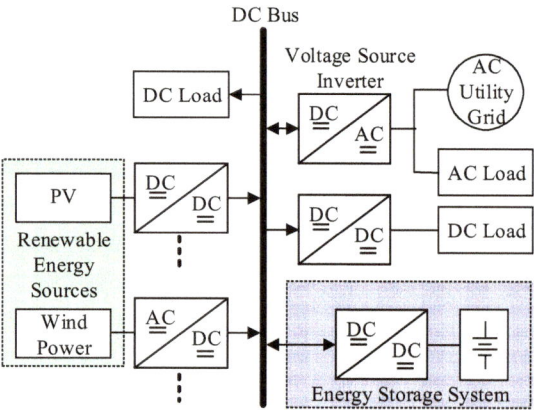

Fig. 1.5 The typical dc microgrid system

cells can be modularly implemented in the power converter stage with much easier scalability regarding power and voltage ratings. In such a way, the inherent dc-fault blocking capability can be acquired naturally since the intermediary ac power stage is embedded in the isolated dc-dc stage [27, 28]. Moreover, the intermediary transformer provides high flexibility for connecting two dc buses with large voltage differences. Therefore, isolated dc-dc converters can serve as universal solutions for applications covering various voltage ratings.

Based on different criterion, isolated dc-dc converters can be classified into different types such as resonant/non-resonant dc-dc converters [29, 30], voltage-source/current-source/impedance-source dc-dc converters [31], etc. In recent years, non-resonant isolated dc-dc converters have drawn some attention in both academic and industry, e.g. full bridge dc-dc converter [32], DAB dc-dc converter [1], and their variant topologies [33–36]. Without losing generality, the simplified circuit of one power-transferred branch in the above dc-dc converters can be modeled universally as shown in Fig. 1.6, which indicates a strong dynamic equivalence among these converters featuring I^2ACL configuration. Compared

Fig. 1.6 The simplified circuit of one power transferring branch in the I^2ACL isolated dc-dc converter

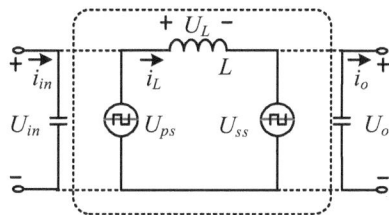

with the DAB dc-dc converter, the research of advanced dynamic control schemes for the other I^2ACL dc-dc converters is not mature. Moreover, up to now, the dynamic equivalence among these isolated dc-dc converters has not been discovered, let alone the unified fast-dynamic control method.

1.2 The IPOP and IIOP DAB DC-DC Converter Systems

In this section, the applications of the IPOP DAB dc-dc converter system and the IIOP DAB dc-dc converter system are discussed separately. Moreover, the existing dynamic control schemes for the IPOP DAB dc-dc converter system are reviewed, and the drawbacks of these schemes are analyzed. Besides, the requirement of hot swap operation are discussed. In addition, the existing dynamic control schemes for the IIOP DAB dc-dc converter are reviewed, and the drawbacks of these schemes are analyzed. Since the IIOP DAB dc-dc converter is employed to connect the multiple power sources to the common dc voltage terminal, the requirement of independent control of each DAB module is analyzed for flexible hot-swap operations with uninterrupted power supply.

1.2.1 The IPOP DAB DC-DC Converter System

Currently, the DAB dc-dc converter is very popular in high-power isolated power conversion applications. Especially for high current and high-power applications, low power DAB dc-dc converters can be flexibly connected in parallel to meet the demand of high-power load [37, 38], and the IPOP DAB dc-dc converter can be shown in Fig. 1.7.

Similar to other output-parallel converter system, it is critical for IPOP DAB dc-dc converters to determine the transferred power of each DAB dc-dc converter flexibly for desired power sharing performance [26, 37–39]. To realize the power sharing control when output terminals of converters are connected in parallel, the droop control strategies are the widely applied technologies in distribution systems [40–41]. However, droop controls always need current sensors for each converter branch to ensure accurate current or power sharing performance, which will result in high cost for converter system with large numbers of the DAB dc-dc modules [42, 43]. To omit the current sensor for each

Fig. 1.7 The topology of
IPOP DAB dc-dc converter
system

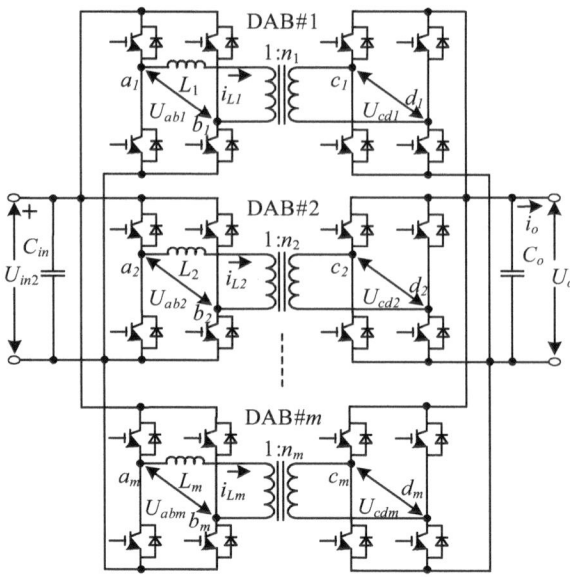

dc-dc converter module, a common-duty-ratio concept is introduced to output-parallel dc-dc converter for passive power sharing control [38, 44], where circuit parameters for each dc-dc converter are regarded as the same. As analyzed in these studies, when parameter mismatches among different converter converters are more than 10%, the current or power sharing performance becomes poor [43, 45].

A current sensorless parameter estimation with current sharing strategy is proposed for the output-parallel DAB dc-dc converter [45]. The core principle of this method is reducing phase-shift ratio of one DAB module to a certain value and using another DAB converter to compensate. In such a way, relationships among circuit parameters of different DAB dc-dc converters can be obtained, and the current or power sharing performance can be acquired. However, the disturbances of the output voltage are inevitable during the estimating process since this method relies on the system's transient information. Moreover, this estimating method is also sensitive to the variation of load resistor because load current is not measured as the feedback value to track the expected required compensation value of transferred power. On the other hand, without current sensor for each DAB module, a model predictive control with power self-balancing performance for the IPOP DAB dc-dc converter system is proposed for power balancing control [46]. However, since circuit parameters including inductance and transformer turn ratio are used to calculate control values, the power balancing performance is sensitive to the variation of these circuit parameters. In addition, the hot swap (plug-in and plug-out) performance is important for the converter system with multiple converter modules [47–49]. Similarly, the plug-in and plug-out control for the IPOP DAB dc-dc converter should be discussed.

1.2.2 The IIOP DAB DC-DC Converter System

In recent years, there is a rapid development of renewable energy system such as pho-tovoltaics (PV), wind turbine (WT) energy and fuel cell, to reduce the reliance on fossil fuels [50–54]. Since these power sources are usually generated as dc power before trans-mission, dc microgrids is currently considered to be an efficient method for integrating distributed renewable resources with less power conversion stages and without traditional issues such as harmonics, synchronization, and unbalance [55–57].

To guarantee the reliable operation and power quality of the microgrid, it is important to mitigate the power fluctuation caused by these renewable energy sources and provide a stable dc-bus voltage. Therefore, the energy storage system (ESS) is usually an indis-pensable part of the dc microgrid to balance the power flowing between the renewable energy source and the load system [58–61]. The typical configuration of the dc micro-grid with ESS can be shown in Fig. 1.8, where the ESS is usually based on multiple energy storage units (ESUs) [60–63]. For the ESS, there are always two main objectives including maintaining the dc-bus voltage and configuring the power sharing performance of different ESUs.

In the islanded dc microgrid with ESS, most research focuses on the traditional dc-dc converters such as buck and boost converters for realizing the bidirectional power trans-mission between dc grid bus and ESS and maintaining the dc grid voltage under different transient conditions [64–66]. However, these traditional dc-dc converters cannot provide electric isolation. Currently, the DAB dc-dc converter with the symmetric, isolated, and

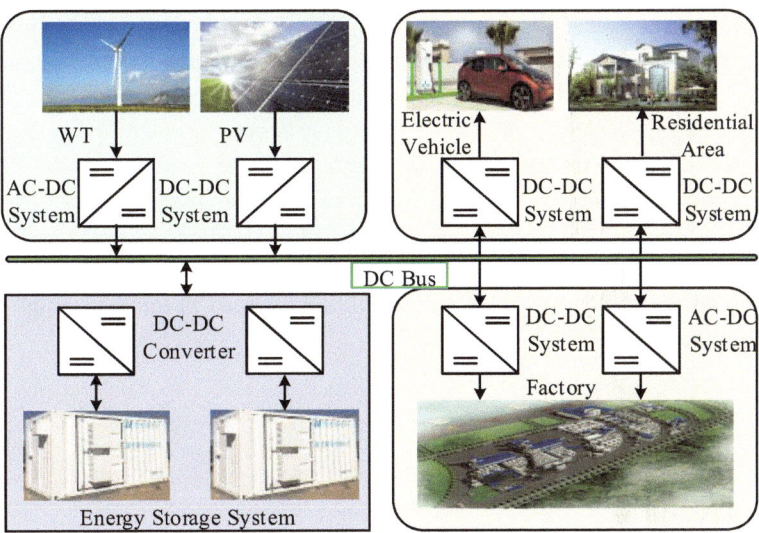

Fig. 1.8 The isolated microgrid with the ESS established by multiple ESUs

bidirectional characteristics becomes a promising candidate for the dc power system [1, 5, 22], which can form cascading or paralleling configurations for different voltage-level requirements. Since the soft switching performance can also be easily implemented, the high efficiency and high-power density are the advantages of this converter. Moreover, the ultrafast dynamic performance under input-voltage or load disturbances of DAB dc-dc converter can be very easily achieved, which can boost the robustness of the dc microgrid [67–69].

The IIOP DAB dc-dc converter system can be shown in Fig. 1.9, which can be employed to connect the multiple energy sources and the dc-link terminal. For the parallel DAB dc-dc converter system, centralized optimized dynamic control strategies with one centralized proportional-integral (PI) controller have been proposed [25, 26, 46], where the fast-dynamic performance can be provided for ensuring a strong dc-bus voltage. However, when a new DAB-based ESU should be plugged in for extending the power capacity of the ESS, the reprogramming operation is unavoidable with one centralized PI control structure, which is not suitable for the islanded dc microgrid. For this condition, the droop control concept may be a promising candidate. When the steady-state condition of the isolated microgrid system is achieved, the power sharing performance of the energy storage system can be determined by the droop coefficients [70, 71]. Sometimes, the power sharing performance under the droop concept is degraded by the line impedance since the measured dc bus voltages for different energy storage units may be different caused by the line impedance. Therefore, an accurate power sharing control method is proposed to reduce the impact of the line resistance by adding the line resistor in the droop control structure [72]. Nevertheless, the accurate line resistors may be difficult to obtain in practical application, and these line resistors are changed with the temperature and the network structure of the power system. Further, an improved droop control method with low bandwidth communication is proposed to detect the actual output voltage of each energy storage module and adjust the droop coefficient for accurate current sharing performance [73]. Then, when the centralized energy storage system is adopted, the line impedance can be neglectful, and with suitable droop coefficients, the accurate power sharing performance among different energy storage units can be acquired. Therefore, the study of droop control concept should be presented for the IIOP DAB dc-dc converter system with decentralized control system for flexible hot-swap operations with uninterrupted power supply, especially when a new DAB unit should be plugged in this converter system for higher power capacity.

Fig. 1.9 The topology of the IIOP DAB dc-dc converter system

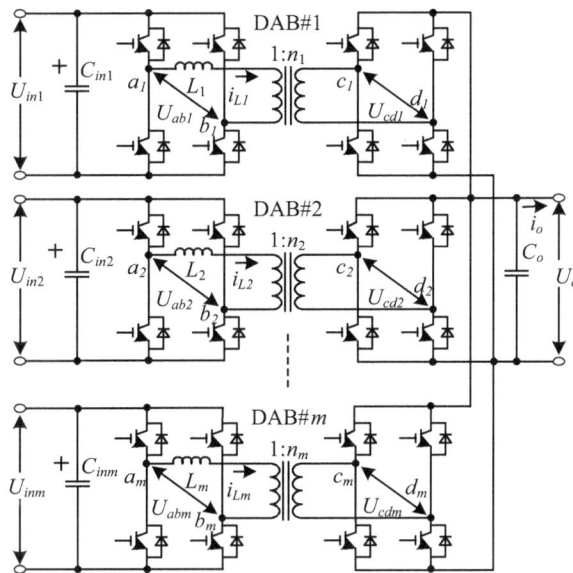

1.3   The IPOS and ISOP DAB DC-DC Converter Systems

In this section, the applications of the IPOS DAB dc-dc converter system and the ISOP DAB dc-dc converter system are discussed separately. Moreover, the existing dynamic control schemes for the IPOS DAB dc-dc converter system are reviewed, and the drawbacks of these schemes are analyzed. In addition, the existing dynamic control schemes for the ISOP DAB dc-dc converter are reviewed, and the drawbacks of these schemes are analyzed. Furthermore, to ensure the accuracy of the desired power sharing performance, the inductance estimating scheme should be further studied for the ISOP DAB dc-dc converter system.

1.3.1   The IPOS DAB DC-DC Converter System

Along with the development of renewable energies such as wind turbine energy and solar energy [74], the dc power distribution system has been a promising alternative to collect, transfer and distribute these energies. Since these power sources always generate unstable electrical power, the ESS system has become an essential technology to boost robustness and stability of dc power distribution system [75]. ESS components such as batteries and super capacitors always offer low output voltage. Therefore, the output-series dc-dc converters with independent or common input terminals and modular multilevel converter isolated dc-dc converter system can be employed to connect the low voltage dc (LVDC) component and the medium voltage dc (MVDC) bus [76–78].

For the output-series dc-dc converter, some strategies are presented to deal with the voltage sharing performance for the input-series structure [79, 80]. These strategies usually focus on the input side of converter system, and the output voltage sharing is naturally allocated by obeying the Conservation of Energy Principle. Therefore, these schemes are not completely suitable for the output-series dc-dc converters with independent or common input terminals. Moreover, based on the PI controller, some methods are presented to address the output voltage sharing performance for input-parallel-output-series dc-dc converter [81–83], but these strategies restrict the dynamic responses of the output-series dc-dc converter system without accurate adjustment of capacitor voltages.

Since the IPOS DAB dc-dc converter system is adopted widely to link LVDC bus and MVDC bus in dc power system [84, 85], the IPOS DAB dc-dc converter as shown in Fig. 1.10 is studied in this work, where the input side of each DAB module can also be independent for connecting different energy sources. Generally, in dc power system, the IPOS DAB dc-dc converter system with ESS system should meet various controlling requirements, such as:

(1) The uninterrupted power supply for maintaining the output dc-link voltage.
(2) The tunable power sharing ability for state-of-charge-balancing control of different ESS equipment.
(3) The hot swap control of multiple ESS systems for maintenance and replacement of ESS equipment.

Fig. 1.10 The topology of the IPOS DAB dc-dc converter system

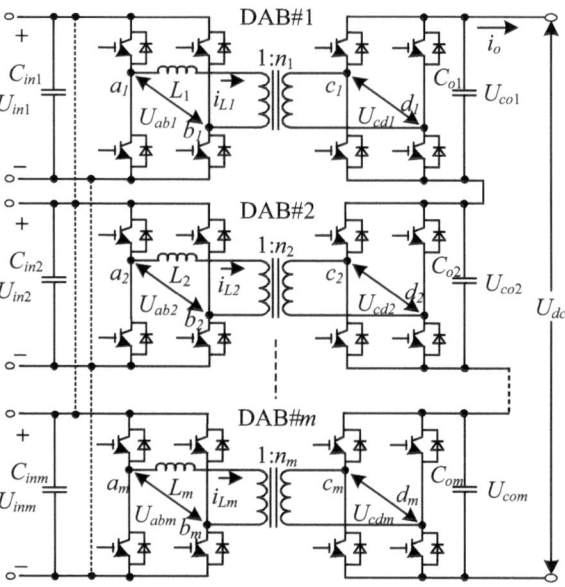

(4) The black-start performance for reducing fluctuations of output capacitor voltages during start-up process.

1.3.2 The ISOP DAB DC-DC Converter System

High-voltage dc converter systems with large voltage turn ratio are extensively used in some power electronic applications such as rail transit system, energy storage system, and microgrids [26, 86–89]. Based on ISOP structure, the power electronic transformers with electric isolation become a promising candidate for connecting the MVDC and the LVDC in these converter systems [90, 91]. Generally, compared with other isolated dc-dc converters, the DAB is more suitable for high power application with high efficiency, bidirectional operation, and fault isolation [1, 5]. So, this work focuses on the ISOP DAB dc-dc converter as shown in Fig. 1.11 for providing stable power to the LVDC side.

With equivalent power sharing control for ISOP converter system, the equivalent utilization of components can be ensured, and the over-voltage over-current issues can be avoided for each converter module [92]. Traditionally, the common duty-ratio operation can be employed to realize power balance performance in parallel/series dc-dc converter system [93, 94], which can significantly reduce the design cost of the controller system. However, the power balance performance is sensitive to the parameter mismatch, especially for the ISOP DAB dc-dc converter system, and the transient process may result in instability under common duty ratio control [38, 95].

Fig. 1.11 The topology of the ISOP DAB dc-dc converter system

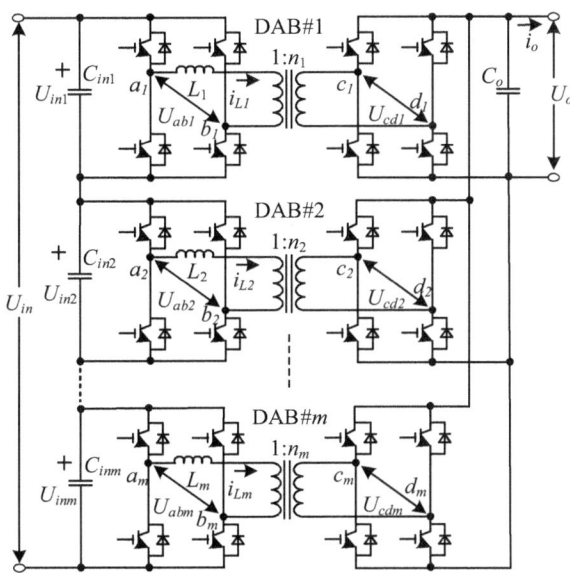

To realize positive power sharing control operation, there are two main ways to realize the power balance control for the ISOP dc-dc converter including the positive input-voltage control scheme [96–99] and the positive transferred-current control scheme [40, 100–103]. For the positive input-voltage control scheme, these existing control strategies can be divided into two groups including the input voltage direct control [96, 97] and the input-voltage droop control [98, 99]. Moreover, for the positive transferred-current control scheme, there are also two categories including the transferred-current direct control [100] and the transferred-current droop control [40, 103]. In [96], a simple sensorless current mode control scheme is proposed for guaranteeing stable sharing performance of the ISOP dc-dc converter system. Similarly, based on the same outer control structure, an input voltage sharing control method is proposed for the ISOP forward dc-dc converter system [97], and the equivalent input voltages can be obtained. For modular converter systems, the droop control concept is also a potential candidate for realizing the power balance performance. Based on the droop structure, a wireless input voltage sharing control method is proposed for the isolated dc-dc converter [98], and a similar decentralized control method is proposed for ISOP DAB dc-dc converter system. Moreover, by directly controlling the transferred current, the power balance operation can be realized, but the converter system prefers to be unstable with the negative resistance model [104]. So, a transferred-current differential control scheme is proposed to address this issue [100]. Furthermore, the transferred-current droop control method can be employed to achieve the power balance control for ISOP dc-dc converter [103]. A decentralized inverse-droop control method is proposed for balancing power sharing of the ISOP isolated dc-dc converter [40].

Since the transferred current of the DAB dc-dc converter contains the ac current, the transferred-current-based control would be not suitable. Moreover, the droop-based control scheme usually results in poor dynamic and steady-state performances [105]. So, the input-voltage direct control may be the most suitable control for the ISOP DAB dc-dc converter [96, 97], but the decoupling between the regulation of input voltage and the adjustment of output voltage in the traditional way may result in a bad transient process. So, a novel input voltage sharing control is proposed to decouple with the output voltage regulation as shown in Fig. 1.12 [106, 107]. However, since the PI controller for adjusting the input voltage of the first two modules is employed to determine the phase-shift ratio not the transferred power, the influence on the output voltage is also obvious. In addition, these existing strategies are more focused on the power sharing performance, and the fast-dynamic response for the ISOP DAB dc-dc converter is not studied for dealing with the disturbances of the input voltage and the load current. Furthermore, as discussion for the IPOP DAB dc-dc converter, the inductance-estimating method should also be studied for ensuring the power sharing performance of the ISOP DAB dc-dc converter.

Fig. 1.12 The traditional decoupling compensation structure for ISOP DAB dc-dc converter system with three modules

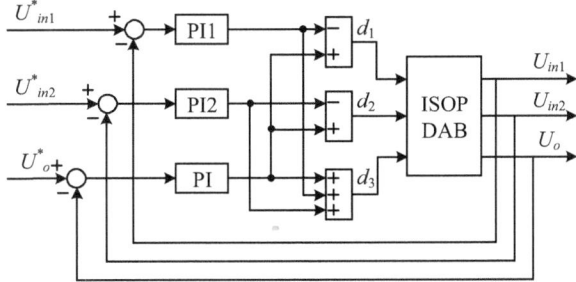

1.4   The DAB-Based PPP Converter System with Robust DC-Link Voltage

With the appeal of carbon neutrality, the installation of renewable energy systems has steadily increased over recent years [50]. Among the renewable energy sources such as photovoltaic (PV) and fuel cell, PV energy has become one of the most important energy sources, especially for the residential PV grid-tied system [51, 52], the railway electrification system [108], and the electric vehicles charger system [109]. In most applications, PV panels are connected in series for achieving higher voltages as shown in Fig. 1.13a [110, 111]. However, because of some undesired factors such as manufacturing tolerances, partial shading and nonuniform aging, the caused mismatch in the PV cells will restrict the total output power of these current-sharing panels.

Although the bypass diodes as shown in Fig. 1.13a can reduce the loss of the output power, the power losses are still high without a positive control, especially for the series-connected PV panels since the total available power of one cell-string may be bypassed for a small difference [112]. Moreover, with bypass diodes, the power-voltage curve will

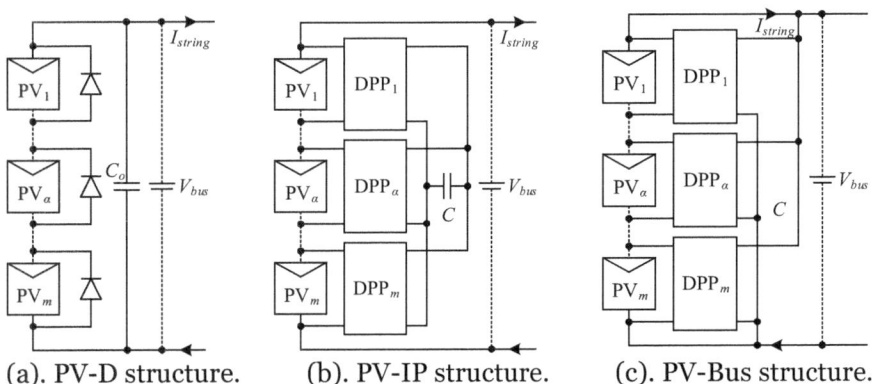

(a). PV-D structure.   (b). PV-IP structure.   (c). PV-Bus structure.

Fig. 1.13 The existing distributed PV architectures for MPPT performance. **a** PV-D structure. **b** PV-IP structure. **c** PV-Bus structure

become complicated with several peak points, so it will be more difficult to realize the global maximum power transmission for the PV system [113]. To deal with this issue, the differential power processing (DPP) technique is a promising technique for realizing the individual maximum power point tracking (MPPT) control [114–116], where the DPP modules are isolated dc-dc converters. These isolated converters are employed to process only the mismatching power among PV panels under the MPPT controls, which can reduce the power rating of these converters and improve the efficiency of the PV-based system [117]. The typical DPP architectures can be shown in Fig. 1.13b as the PV-to-isolated port (PV-IP) structure and Fig. 1.13c as the PV-bus structure. In the PV-IP architecture, the string current is determined by the output power of the PV panels with MPPT, and the string current of the PV-bus structure can be adjusted flexibly. Thus, by optimizing the string current of PV-bus architecture, the efficiency of the PV-bus architecture can be a little higher than that of the PV-IP structure [17, 118]. For these distributed PV architectures as shown in Fig. 1.13, the total transferred current to the dc-link terminal is only from the PV modules, and under MPPT performance, this current is uncontrollable. Therefore, these PV structures are not suitable for the islanded dc microgrid, where the constant dc-link voltage is required.

Although there are lots of existing literatures focusing on the partial power processing (PPP) converter system for PVs, there seems no existing literature which can provide the constant dc-link voltage with the PPP technique. Therefore, to deal with this issue, a PPP converter system with adjustable dc-link voltage for islanded dc microgrid is proposed as shown in Fig. 1.14, which can also act as an alternative scheme when the electricity consumer loses the support of the strong ac grid system. Different from the PV-bus structure, the proposed PPP converter system adopts an energy storage system such as batteries to control the string current. Moreover, the additional dc-dc converter is in series with the PV panels, which can be employed to adjust the total dc-link voltage, and this additional converter system can be the ISOP dc-dc converter for different applications. Similarly, these adopted dc-dc converters should be isolated in the proposed PPP converter system with adjustable dc-link voltage.

In addition, since the ultrafast dynamic performance under input-voltage or load disturbances of the DAB dc-dc converter can be easily achieved, the proposed PPP converter system with DAB modules can be shown in Fig. 1.15 for the stabilized dc microgrid [67–69], where the PV panels can be replaced by other renewable energy sources with current output and limited voltage regulating requirement such as fuel cell and wind turbine with ac-dc conversion. As shown in Fig. 1.15, the mth DAB module is adopted to adjust the total dc-link voltage, and other DAB modules are employed to realize the independent control of the renewable energy source such as MPPT for the PV panels. Moreover, to boost the dynamic performance of this DAB-based PPP converter system, the high-robustness control strategy should be proposed for maintaining the dc-link voltage when the working condition of the renewable energy source, the voltage of the battery, and the load condition are changed.

Fig. 1.14 The proposed PPP
converter system with robust
dc-link voltage for islanded dc
microgrid

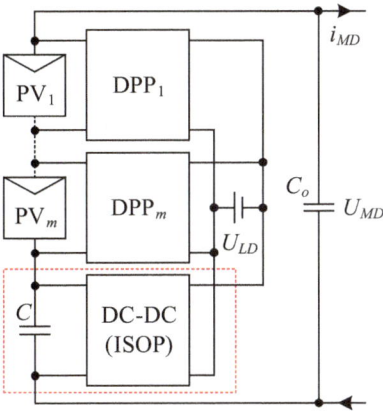

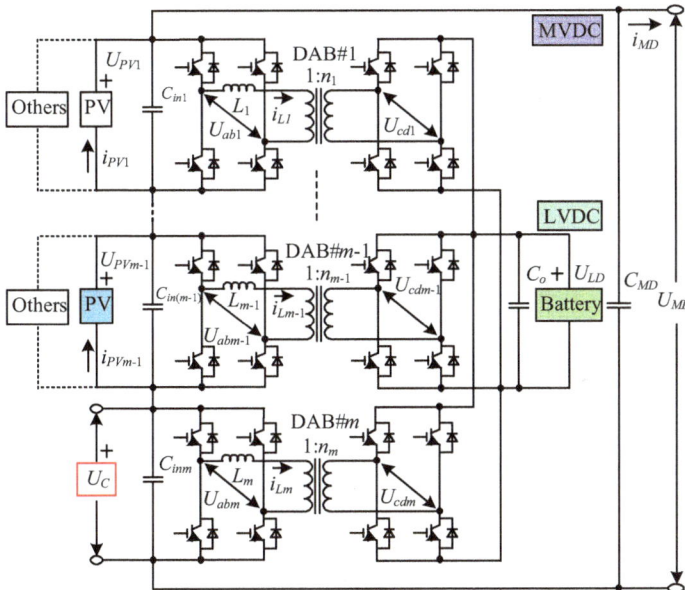

Fig. 1.15 The topology of the proposed DAB-based PPP converter system with adjustable dc-link
voltage

1.5   Research Objectives and Contributions

The DAB dc-dc converter has become a promising candidate for dc-dc applications, so
there are lots of existing literatures which focus on the fast-dynamic response for the
single DAB module in the past thirty years [5, 67, 119, 120]. Meanwhile, some existing
dc-dc converters featuring I^2ACL configuration are also proposed for applications like full

bridge dc-dc converter and three-phase DAB dc-dc converter. However, these I^2ACL dc-dc converters still need further study. Besides, the dynamic equivalence between the DAB dc-dc converter and other I^2ACL dc-dc converters should also be investigated. Moreover, the DAB-based dc-dc converter systems such as IPOP, IIOP, IPOS, and ISOP DAB configurations should be studied, especially in terms of fast-dynamic performances. In addition, the hot-swap operations and the CPE methods should also be investigated for ensuring the performance of these modular DAB dc-dc converter systems. Furthermore, a novel DAB-based PPP converter system is introduced for realizing the independent control of the Renewable energy source and the stabilization of the total dc-link voltage simultaneously. Based on some existing schemes, the requirement of the renewable energy source can be addressed, but the robustness control of the total dc-link voltage should be further studied.

Therefore, to deal with these issues, there are several research objectives and contributions to this work:

(1) Unified Fast-Dynamic Direct-Current Control scheme for I^2ACL DC-DC Converters

In Chap. 2, the existing I^2ACL isolated dc-dc converters are reviewed thoroughly, including unidirectional type and bidirectional type. Besides, the general current transferred features of these two groups are analyzed, respectively. Then, it can be obtained that the transferred current is just influenced by the middle inductance little even during the transient process. So, the I^2ACL isolated dc-dc converter can be regarded as the first-order converter. Based on this characteristic, a unified fast-dynamic direct-current (FDDC) control scheme is proposed for improving the dynamic performance of these I^2ACL isolated dc-dc converters. Such a scheme can also facilitate the uniform control design for existing or emerging new topologies with the same electrical equivalence. Notably, with this fundamental analysis, the control strategies of the DAB-based dc-dc converter systems such as IPOP, IIOP, IPOS, and ISOP DAB dc-dc converter systems can be easily extended to other I^2ACL isolated dc-dc converters with the same configurations.

(2) The flexible power sharing control scheme with the fast-dynamic performance for the IPOP or IIOP DAB dc-dc converter systems

A tunable power sharing strategy with fast-dynamic response is proposed for the IPOP DAB dc-dc converter system in Chap. 3. Based on this tunable power sharing control strategy, excellent dynamic performance under disturbances of the input voltage and the load resistor can be achieved. However, inaccurate circuit-parameter information always damages the power sharing performance among different DAB modules. Therefore, the comprehensive circuit-parameter estimating (CPE) schemes are proposed for different conditions of the IPOP DAB dc-dc converter system including the start-up process, the working process, and the plugging-in operation of a DAB dc-dc converter, respectively.

The proposed CPE methods can be employed in the IIOP DAB dc-dc converter system with centralized controller. Besides, the hot swap operations of the DAB module is discussed. In addition, a communication-free power management strategy is proposed to maintain the dc-link voltage of the IIOP DAB dc-dc converter system with decentralized controllers for each module in Chap. 3. Based on the proposed scheme, the high robustness of the dc-link voltage can be ensured when the input voltage, the load condition, and the power sharing performance are changed. Meanwhile, the proposed strategy ensures seamless plug-in or -out operations of the DAB module with the uninterrupted power supply.

(3) The flexible power sharing control scheme with the fast-dynamic performance for the IPOS or ISOP DAB dc-dc converter systems

A tunable power sharing control strategy of the IPOS DAB dc-dc converter is proposed for maintaining the output voltage and managing the power sharing performance in Chap. 4. Besides, with a small variant, the proposed scheme can realize the black-start operation for this DAB based dc-dc converter system. Based on this variant scheme, the synchronous charging of the output capacitors can be achieved. Moreover, an input-oriented power sharing control scheme with fast-dynamic response is proposed for the ISOP DAB dc-dc converter system in Chap. 4. Compared with existing methods, this proposed scheme can significantly reduce the coupling between the power sharing control and the output voltage regulation. In addition, a general inductance-estimating method is proposed for ensuring the power sharing performance of the ISOP DAB dc-dc converter system. Similarly, the general inductance-estimating method can also be employed in the IPOS DAB dc-dc converter system.

(4) The Partial Power Processing Converter System with Robust DC-Link Voltage for Islanded dc Microgrid

In Chap. 5, based on the proposed DAB-based PPP converter system, the operating principle and simplified circuit of this PPP converter system is discussed first. Moreover, to boost the robustness of the dc-link voltage, a high-robustness control strategy is proposed for maintaining the total dc-link voltage under different cases: (1) The working condition of the renewable energy source is changed. (2) The voltage of the battery is varied. (3) The load condition is changed. In addition, the operation when one renewable energy source is out of work is also presented. Notably, the renewable energy should feature the current output and the limited output-voltage regulation such as PV, fuel cell and WT with ac-dc conversion. By using the PV as an example, the function of the novel DAB-based PPP converter system is investigated, and the effectiveness of the proposed high-robustness control strategy is verified.

References

1. R. W. A. A. D. Doncker, D. M. Divan, and M. H. Kheraluwala, "A three-phase soft-switched high-power-density DC/DC converter for high-power applications," *IEEE Transactions on Industry Applications,* vol. 27, no. 1, pp. 63–73, 1991.

2. B. Zhao, Q. Song, W. Liu, and Y. Xiao, "Next-Generation Multi-Functional Modular Intelligent UPS System for Smart Grid," *IEEE Transactions on Industrial Electronics,* vol. 60, no. 9, pp. 3602–3618, 2013.

3. F. D. Freijedo, E. Rodriguez-Diaz, and D. Dujic, "Stable and Passive High-Power Dual Active Bridge Converters Interfacing MVDC Grids," *IEEE Transactions on Industrial Electronics,* vol. 65, no. 12, pp. 9561–9570, 2018.

4. F. Krismer, and J. W. Kolar, "Efficiency-Optimized High-Current Dual Active Bridge Converter for Automotive Applications," *IEEE Transactions on Industrial Electronics,* vol. 59, no. 7, pp. 2745–2760, 2012.

5. N. Hou, and Y. W. Li, "Overview and Comparison of Modulation and Control Strategies for a Nonresonant Single-Phase Dual-Active-Bridge DC–DC Converter," *IEEE Transactions on Power Electronics,* vol. 35, no. 3, pp. 3148–3172, 2020.

6. F. Xue, R. Yu, and A. Q. Huang, "A 98.3% Efficient GaN Isolated Bidirectional DC–DC Converter for DC Microgrid Energy Storage System Applications," *IEEE Transactions on Industrial Electronics,* vol. 64, no. 11, pp. 9094–9103, 2017.

7. N. M. L. Tan, T. Abe, and H. Akagi, "Design and Performance of a Bidirectional Isolated DC–DC Converter for a Battery Energy Storage System," *IEEE Transactions on Power Electronics,* vol. 27, no. 3, pp. 1237–1248, 2012.

8. V. Blahnik, Z. Peroutka, J. Molnar, and J. Michalik, "Control of primary voltage source active rectifiers for traction converter with medium-frequency transformer," *2008 13th International Power Electronics and Motion Control Conference*, 2008, pp. 1535–1541.

9. R. Suryadevara, and L. Parsa, "Full-Bridge ZCS-Converter-Based High-Gain Modular DC-DC Converter for PV Integration With Medium-Voltage DC Grids," *IEEE Transactions on Energy Conversion,* vol. 34, no. 1, pp. 302–312, 2019.

10. J. Hu, S. Cui, S. Wang, and R. W. D. Doncker, "Instantaneous Flux and Current Control for a Three-Phase Dual-Active Bridge DC–DC Converter," *IEEE Transactions on Power Electronics,* vol. 35, no. 2, pp. 2184–2195, 2020.

11. M. Guan, "A Series-Connected Offshore Wind Farm Based on Modular Dual-Active-Bridge (DAB) Isolated DC–DC Converter," *IEEE Transactions on Energy Conversion,* vol. 34, no. 3, pp. 1422–1431, 2019.

12. Q. Tian, A. Q. Huang, H. Teng, J. Lu, K. H. Bai, A. Brown, and M. McAmmond, "A novel energy balanced variable frequency control for input-series-output-parallel modular EV fast charging stations," *2016 IEEE Energy Conversion Congress and Exposition (ECCE)*, 2016, pp. 1–6.

13. Z. Guo, R. Yu, W. Xu, X. Feng, and A. Q. Huang, "Design and Optimization of a 200-kW Medium-Frequency Transformer for Medium-Voltage SiC PV Inverters," *IEEE Transactions on Power Electronics,* vol. 36, no. 9, pp. 10548–10560, 2021.

14. J. Yao, W. Chen, C. Xue, Y. Yuan, and T. Wang, "An ISOP Hybrid DC Transformer Combining Multiple SRCs and DAB Converters to Interconnect MVDC and LVDC Distribution Networks," *IEEE Transactions on Power Electronics,* vol. 35, no. 11, pp. 11442–11452, 2020.

15. Q. Sun, Y. Li, G. Liu, Y. Wang, J. Meng, and Q. Mu, "Multiple-Modular High-Frequency DC Transformer With Parallel Clamping Switched Capacitor for Flexible MVDC and HVDC System Applications," *IEEE Journal of Emerging and Selected Topics in Power Electronics,* vol. 8, no. 4, pp. 4130–4143, 2020.

16. D. Neumayr, G. C. Knabben, E. Varescon, D. Bortis, and J. W. Kolar, "Comparative Evaluation of a Full- and Partial-Power Processing Active Power Buffer for Ultracompact Single-Phase DC/AC Converter Systems," *IEEE Journal of Emerging and Selected Topics in Power Electronics,* vol. 9, no. 2, pp. 1994–2013, 2021.

17. C. Olalla, C. Deline, D. Clement, Y. Levron, M. Rodriguez, and D. Maksimovic, "Performance of Power-Limited Differential Power Processing Architectures in Mismatched PV Systems," *IEEE Transactions on Power Electronics,* vol. 30, no. 2, pp. 618–631, 2015.

18. H. Cha, R. Ding, Q. Tang, and F. Z. Peng, "Design and Development of High-Power DC–DC Converter for Metro Vehicle System," *IEEE Transactions on Industry Applications,* vol. 44, no. 6, pp. 1795–1804, 2008.

19. H. Plesko, J. Biela, J. Luomi, and J. W. Kolar, "Novel Concepts for Integrating the Electric Drive and Auxiliary DC–DC Converter for Hybrid Vehicles," *IEEE Transactions on Power Electronics,* vol. 23, no. 6, pp. 3025–3034, 2008.

20. N. M. L. Tan, S. Inoue, A. Kobayashi, and H. Akagi, "Voltage Balancing of a 320-V, 12-F Electric Double-Layer Capacitor Bank Combined With a 10-kW Bidirectional Isolated DC–DC Converter," *IEEE Transactions on Power Electronics,* vol. 23, no. 6, pp. 2755–2765, 2008.

21. D. Rothmund, T. Guillod, D. Bortis, and J. W. Kolar, "99% Efficient 10 kV SiC-Based 7 kV/ 400 V DC Transformer for Future Data Centers," *IEEE Journal of Emerging and Selected Topics in Power Electronics,* vol. 7, no. 2, pp. 753–767, 2019.

22. B. Zhao, Q. Song, W. Liu, and Y. Sun, "Overview of Dual-Active-Bridge Isolated Bidirectional DC–DC Converter for High-Frequency-Link Power-Conversion System," *IEEE Transactions on Power Electronics,* vol. 29, no. 8, pp. 4091–4106, 2014.

23. S. Inoue, and H. Akagi, "A Bidirectional Isolated DC–DC Converter as a Core Circuit of the Next-Generation Medium-Voltage Power Conversion System," *IEEE Transactions on Power Electronics,* vol. 22, no. 2, pp. 535–542, 2007.

24. M. Kwon, and S. Choi, "Control Scheme for Autonomous and Smooth Mode Switching of Bidirectional DC–DC Converters in a DC Microgrid," *IEEE Transactions on Power Electronics,* vol. 33, no. 8, pp. 7094–7104, 2018.

25. N. Hou, and Y. Li, "The Comprehensive Circuit-Parameter Estimating Strategies for Output-Parallel Dual-Active-Bridge DC–DC Converters With Tunable Power Sharing Control," *IEEE Transactions on Industrial Electronics,* vol. 67, no. 9, pp. 7583–7594, 2020.

26. J. Liu, J. Yang, J. Zhang, Z. Nan, and Q. Zheng, "Voltage Balance Control Based on Dual Active Bridge DC/DC Converters in a Power Electronic Traction Transformer," *IEEE Transactions on Power Electronics,* vol. 33, no. 2, pp. 1696–1714, 2018.

27. L. Wang, Q. Zhu, W. Yu, and A. Q. Huang, "A Medium-Voltage Medium-Frequency Isolated DC–DC Converter Based on 15-kV SiC MOSFETs," *IEEE Journal of Emerging and Selected Topics in Power Electronics,* vol. 5, no. 1, pp. 100–109, 2017.

28. H. Tarzamni, F. P. Esmaeelnia, M. Fotuhi-Firuzabad, F. Tahami, S. Tohidi, and P. Dehghanian, "Comprehensive Analytics for Reliability Evaluation of Conventional Isolated Multiswitch PWM DC–DC Converters," *IEEE Transactions on Power Electronics,* vol. 35, no. 5, pp. 5254–5266, 2020.

29. S. A. Gorji, H. G. Sahebi, M. Ektesabi, and A. B. Rad, "Topologies and Control Schemes of Bidirectional DC–DC Power Converters: An Overview," *IEEE Access,* vol. 7, pp. 117997–118019, 2019.

30. X. Pan, H. Li, Y. Liu, T. Zhao, C. Ju, and A. K. Rathore, "An Overview and Comprehensive Comparative Evaluation of Current-Fed-Isolated-Bidirectional DC/DC Converter," *IEEE Transactions on Power Electronics,* vol. 35, no. 3, pp. 2737–2763, 2020.

31. A. Chub, D. Vinnikov, F. Blaabjerg, and F. Z. Peng, "A Review of Galvanically Isolated Impedance-Source DC–DC Converters," *IEEE Transactions on Power Electronics,* vol. 31, no. 4, pp. 2808–2828, 2016.

32. Y. Xie, R. Ghaemi, J. Sun, and J. S. Freudenberg, "Implicit Model Predictive Control of a Full Bridge DC–DC Converter," *IEEE Transactions on Power Electronics,* vol. 24, no. 12, pp. 2704–2713, 2009.

33. A. J. B. Bottion, and I. Barbi, "Input-Series and Output-Series Connected Modular Output Capacitor Full-Bridge PWM DC–DC Converter," *IEEE Transactions on Industrial Electronics,* vol. 62, no. 10, pp. 6213–6221, 2015.

34. C. Zhao, S. D. Round, and J. W. Kolar, "An Isolated Three-Port Bidirectional DC-DC Converter With Decoupled Power Flow Management," *IEEE Transactions on Power Electronics,* vol. 23, no. 5, pp. 2443–2453, 2008.

35. L. Wang, D. Zhang, Y. Wang, B. Wu, and H. S. Athab, "Power and Voltage Balance Control of a Novel Three-Phase Solid-State Transformer Using Multilevel Cascaded H-Bridge Inverters for Microgrid Applications," *IEEE Transactions on Power Electronics,* vol. 31, no. 4, pp. 3289–3301, 2016.

36. J. Hu, Z. Yang, S. Cui, and R. W. D. Doncker, "Closed-Form Asymmetrical Duty-Cycle Control to Extend the Soft-Switching Range of Three-Phase Dual-Active-Bridge Converters," *IEEE Transactions on Power Electronics,* vol. 36, no. 8, pp. 9609–9622, 2021.

37. R. Haneda, and H. Akagi, "Experimental Characterization of a 750-V 100-kW 16-kHz Bidirectional Isolated DC-DC Converter With a Unity-Turns-Ratio Transformer at Different Voltage Ratios," *2020 IEEE Energy Conversion Congress and Exposition (ECCE),* 2020, pp. 649–655.

38. J. Shi, L. Zhou, and X. He, "Common-Duty-Ratio Control of Input-Parallel Output-Parallel (IPOP) Connected DC–DC Converter Modules With Automatic Sharing of Currents," *IEEE Transactions on Power Electronics,* vol. 27, no. 7, pp. 3277–3291, 2012.

39. M. M. U. Rehman, F. Zhang, R. Zane, and D. Maksimovic, "Control of bidirectional DC/DC converters in reconfigurable, modular battery systems," *2017 IEEE Applied Power Electronics Conference and Exposition (APEC),* 2017, pp. 1277–1283.

40. Y. Zhang, and Y. W. Li, "Energy Management Strategy for Supercapacitor in Droop-Controlled DC Microgrid Using Virtual Impedance," *IEEE Transactions on Power Electronics,* vol. 32, no. 4, pp. 2704–2716, 2017.

41. J. He, Y. Li, C. Wang, Y. Pan, C. Zhang, and X. Xing, "Hybrid Microgrid With Parallel- and Series-Connected Microconverters," *IEEE Transactions on Power Electronics,* vol. 33, no. 6, pp. 4817–4831, 2018.

42. Y. Lin, Y. Wang, S. Wang, and H. Li, "Sensorless current estimation and sharing in multiphase input-parallel output-parallel DC-DC converters." pp. 1–6.

43. W. Shanshan, W. Yubin, and L. Yifei, "Sensorless power balance control for cascaded multilevel converter based solid state transformer," *2015 IEEE 2nd International Future Energy Electronics Conference (IFEEC),* 2015, pp. 1–6.

44. J. Shi, T. Liu, J. Cheng, and X. He, "Automatic Current Sharing of an Input-Parallel Output-Parallel (IPOP)-Connected DC–DC Converter System With Chain-Connected Rectifiers," *IEEE Transactions on Power Electronics,* vol. 30, no. 6, pp. 2997–3016, 2015.

45. Y. Wang, F. Wang, Y. Lin, and T. Hao, "Sensorless parameter estimation and current-sharing strategy in two-phase and multiphase IPOP DAB DC–DC converters," *IET Power Electronics,* vol. 11, no. 6, pp. 1135–1142, 2018.

46. F. An, W. Song, B. Yu, and K. Yang, "Model Predictive Control With Power Self-Balancing of the Output Parallel DAB DC–DC Converters in Power Electronic Traction Transformer," *IEEE Journal of Emerging and Selected Topics in Power Electronics,* vol. 6, no. 4, pp. 1806–1818, 2018.

47. Y. Han, K. Zhang, H. Li, E. A. A. Coelho, and J. M. Guerrero, "MAS-Based Distributed Coordinated Control and Optimization in Microgrid and Microgrid Clusters: A Comprehensive Overview," *IEEE Transactions on Power Electronics,* vol. 33, no. 8, pp. 6488–6508, 2018.

48. Y. Ota, H. Taniguchi, T. Nakajima, K. M. Liyanage, J. Baba, and A. Yokoyama, "Autonomous Distributed V2G (Vehicle-to-Grid) Satisfying Scheduled Charging," *IEEE Transactions on Smart Grid,* vol. 3, no. 1, pp. 559–564, 2012.

49. N. F. Avila, and C. Chu, "Distributed Pinning Droop Control in Isolated AC Microgrids," *IEEE Transactions on Industry Applications,* vol. 53, no. 4, pp. 3237–3249, 2017.

50. B. Liu, F. Zhuo, Y. Zhu, and H. Yi, "System Operation and Energy Management of a Renewable Energy-Based DC Micro-Grid for High Penetration Depth Application," *IEEE Transactions on Smart Grid,* vol. 6, no. 3, pp. 1147–1155, 2015.

51. L. Yunwei, D. M. Vilathgamuwa, and L. Poh Chiang, "Design, analysis, and real-time testing of a controller for multibus microgrid system," *IEEE Transactions on Power Electronics,* vol. 19, no. 5, pp. 1195–1204, 2004.

52. X. Li, L. Guo, C. Hong, Y. Zhang, Y. W. Li, and C. Wang, "Hierarchical Control of Multiterminal DC Grids for Large-Scale Renewable Energy Integration," *IEEE Transactions on Sustainable Energy,* vol. 9, no. 3, pp. 1448–1457, 2018.

53. N. L. Díaz, A. C. Luna, J. C. Vasquez, and J. M. Guerrero, "Centralized Control Architecture for Coordination of Distributed Renewable Generation and Energy Storage in Islanded AC Microgrids," *IEEE Transactions on Power Electronics,* vol. 32, no. 7, pp. 5202–5213, 2017.

54. T. Kovaltchouk, A. Blavette, J. Aubry, H. B. Ahmed, and B. Multon, "Comparison Between Centralized and Decentralized Storage Energy Management for Direct Wave Energy Converter Farm," *IEEE Transactions on Energy Conversion,* vol. 31, no. 3, pp. 1051–1058, 2016.

55. P. Karlsson, and J. Svensson, "DC bus voltage control for a distributed power system," *IEEE Transactions on Power Electronics,* vol. 18, no. 6, pp. 1405–1412, 2003.

56. K. Sun, L. Zhang, Y. Xing, and J. M. Guerrero, "A Distributed Control Strategy Based on DC Bus Signaling for Modular Photovoltaic Generation Systems With Battery Energy Storage," *IEEE Transactions on Power Electronics,* vol. 26, no. 10, pp. 3032–3045, 2011.

57. N. Hou, and Y. W. Li, "A Tunable Power Sharing Control Scheme for the Output-Series DAB DC–DC System With Independent or Common Input Terminals," *IEEE Transactions on Power Electronics,* vol. 34, no. 10, pp. 9386–9391, 2019.

58. X. Shen, M. Shahidehpour, Y. Han, S. Zhu, and J. Zheng, "Expansion Planning of Active Distribution Networks With Centralized and Distributed Energy Storage Systems," *IEEE Transactions on Sustainable Energy,* vol. 8, no. 1, pp. 126–134, 2017.

59. J. Xiao, P. Wang, and L. Setyawan, "Hierarchical Control of Hybrid Energy Storage System in DC Microgrids," *IEEE Transactions on Industrial Electronics,* vol. 62, no. 8, pp. 4915–4924, 2015.

60. C. Yuan, M. A. Haj-ahmed, and M. S. Illindala, "Protection Strategies for Medium-Voltage Direct-Current Microgrid at a Remote Area Mine Site," *IEEE Transactions on Industry Applications,* vol. 51, no. 4, pp. 2846–2853, 2015.

61. M. B. Shadmand, and R. S. Balog, "Multi-Objective Optimization and Design of Photovoltaic-Wind Hybrid System for Community Smart DC Microgrid," *IEEE Transactions on Smart Grid,* vol. 5, no. 5, pp. 2635–2643, 2014.

62. P. García, P. Arboleya, B. Mohamed, and A. A. C. Vega, "Implementation of a Hybrid Distributed/Centralized Real-Time Monitoring System for a DC/AC Microgrid With Energy Storage Capabilities," *IEEE Transactions on Industrial Informatics,* vol. 12, no. 5, pp. 1900–1909, 2016.

63. A. Agrawal, C. S. Nalamati, and R. Gupta, "Hybrid DC–AC Zonal Microgrid Enabled by Solid-State Transformer and Centralized ESD Integration," *IEEE Transactions on Industrial Electronics,* vol. 66, no. 11, pp. 9097–9107, 2019.

64. J. Xiao, P. Wang, and L. Setyawan, "Multilevel Energy Management System for Hybridization of Energy Storages in DC Microgrids," *IEEE Transactions on Smart Grid,* vol. 7, no. 2, pp. 847–856, 2016.

65. T. R. Oliveira, W. W. A. G. Silva, and P. F. Donoso-Garcia, "Distributed Secondary Level Control for Energy Storage Management in DC Microgrids," *IEEE Transactions on Smart Grid,* vol. 8, no. 6, pp. 2597–2607, 2017.

66. Y. Xia, M. Yu, P. Yang, Y. Peng, and W. Wei, "Generation-Storage Coordination for Islanded DC Microgrids Dominated by PV Generators," *IEEE Transactions on Energy Conversion,* vol. 34, no. 1, pp. 130–138, 2019

67. F. An, W. Song, K. Yang, N. Hou, and J. Ma, "Improved dynamic performance of dual active bridge dc–dc converters using MPC scheme," *IET Power Electronics,* vol. 11, no. 11, pp. 1756–1765, 2018

68. W. Song, N. Hou, and M. Wu, "Virtual Direct Power Control Scheme of Dual Active Bridge DC–DC Converters for Fast Dynamic Response," *IEEE Transactions on Power Electronics,* vol. 33, no. 2, pp. 1750–1759, 2018.

69. G. G. Oggier, M. Ordonez, J. M. Galvez, and F. Luchino, "Fast Transient Boundary Control and Steady-State Operation of the Dual Active Bridge Converter Using the Natural Switching Surface," *IEEE Transactions on Power Electronics,* vol. 29, no. 2, pp. 946–957, 2014.

70. R. Majumder, B. Chaudhuri, A. Ghosh, R. Majumder, G. Ledwich, and F. Zare, "Improvement of Stability and Load Sharing in an Autonomous Microgrid Using Supplementary Droop Control Loop," *IEEE Transactions on Power Systems,* vol. 25, no. 2, pp. 796–808, 2010

71. X. Li, L. Guo, S. Zhang, C. Wang, Y. W. Li, A. Chen, and Y. Feng, "Observer-Based DC Voltage Droop and Current Feed-Forward Control of a DC Microgrid," *IEEE Transactions on Smart Grid,* vol. 9, no. 5, pp. 5207–5216, 2018.

72. K. D. Hoang, and H. Lee, "Accurate Power Sharing With Balanced Battery State of Charge in Distributed DC Microgrid," *IEEE Transactions on Industrial Electronics,* vol. 66, no. 3, pp. 1883–1893, 2019.

73. X. Lu, J. M. Guerrero, K. Sun, and J. C. Vasquez, "An Improved Droop Control Method for DC Microgrids Based on Low Bandwidth Communication With DC Bus Voltage Restoration and Enhanced Current Sharing Accuracy," *IEEE Transactions on Power Electronics,* vol. 29, no. 4, pp. 1800–1812, 2014.

74. F. Blaabjerg, R. Teodorescu, M. Liserre, and A. V. Timbus, "Overview of Control and Grid Synchronization for Distributed Power Generation Systems," *IEEE Transactions on Industrial Electronics,* vol. 53, no. 5, pp. 1398–1409, 2006.

75. F. Nejabatkhah, and Y. W. Li, "Overview of Power Management Strategies of Hybrid AC/DC Microgrid," *IEEE Transactions on Power Electronics,* vol. 30, no. 12, pp. 7072–7089, 2015.

76. Q. Song, B. Zhao, J. Li, and W. Liu, "An Improved DC Solid State Transformer Based on Switched Capacitor and Multiple-Phase-Shift Shoot-Through Modulation for Integration of LVDC Energy Storage System and MVDC Distribution Grid," *IEEE Transactions on Industrial Electronics,* vol. 65, no. 8, pp. 6719–6729, 2018.

77. I. A. Gowaid, G. P. Adam, A. M. Massoud, S. Ahmed, and B. W. Williams, "Hybrid and Modular Multilevel Converter Designs for Isolated HVDC–DC Converters," *IEEE Journal of Emerging and Selected Topics in Power Electronics,* vol. 6, no. 1, pp. 188–202, 2018.

78. Z. Xing, X. Ruan, H. You, X. Yang, D. Yao, and C. Yuan, "Soft-Switching Operation of Isolated Modular DC/DC Converters for Application in HVDC Grids," *IEEE Transactions on Power Electronics,* vol. 31, no. 4, pp. 2753–2766, 2016.

79. L. Qu, D. Zhang, and Z. Bao, "Active Output-Voltage-Sharing Control Scheme for Input Series Output Series Connected DC–DC Converters Based on a Master Slave Structure," *IEEE Transactions on Power Electronics,* vol. 32, no. 8, pp. 6638–6651, 2017.

80. Q. Wei, B. Wu, D. Xu, and N. R. Zargari, "Model Predictive Control of Capacitor Voltage Balancing for Cascaded Modular DC–DC Converters," *IEEE Transactions on Power Electronics,* vol. 32, no. 1, pp. 752–761, 2017.

81. A. Darwish, D. Holliday, and S. Finney, "Operation and control design of an input-series–input-parallel–output-series conversion scheme for offshore DC wind systems," *IET Power Electronics,* vol. 10, no. 15, pp. 2092–2103, 2017.

82. K. G. Anjana, M. Aniruddha Kamath, and M. Barai, "A differential current compensation technique for PV systems under partially shaded condition," *2017 11th IEEE International Conference on Compatibility, Power Electronics and Power Engineering (CPE-POWERENG),* 2017, pp. 116–120.

83. S. Lee, Y. Jeung, and D. Lee, "Output voltage regulation of IPOS modular dual active bridge DC/DC converters using sliding mode control," *2018 IEEE Applied Power Electronics Conference and Exposition (APEC),* 2018, pp. 3062–3067.

84. M. Stieneker, and R. W. D. Doncker, "Dual-active bridge dc-dc converter systems for medium-voltage DC distribution grids," *2015 IEEE 13th Brazilian Power Electronics Conference and 1st Southern Power Electronics Conference (COBEP/SPEC),* 2015, pp. 1–6.

85. T. Todorčević, R. v. Kessel, P. Bauer, and J. A. Ferreira, "A Modulation Strategy for Wide Voltage Output in DAB-Based DC–DC Modular Multilevel Converter for DEAP Wave Energy Conversion," *IEEE Journal of Emerging and Selected Topics in Power Electronics,* vol. 3, no. 4, pp. 1171–1181, 2015.

86. K. Jung-Won, Y. Jung-Sik, and B. H. Cho, "Modeling, control, and design of input-series-output-parallel-connected converter for high-speed-train power system," *IEEE Transactions on Industrial Electronics,* vol. 48, no. 3, pp. 536–544, 2001.

87. J. Duan, D. Zhang, L. Wang, Z. Zhou, and Y. Gu, "Active Voltage Sharing Module for Input-Series Connected Modular DC/DC Converters," *IEEE Transactions on Power Electronics,* vol. 35, no. 6, pp. 5987–6000, 2020.

88. D. Ma, W. Chen, and X. Ruan, "A Review of Voltage/Current Sharing Techniques for Series–Parallel-Connected Modular Power Conversion Systems," *IEEE Transactions on Power Electronics,* vol. 35, no. 11, pp. 12383–12400, 2020.

89. J. Xu, C. Gao, J. Ding, X. Shi, M. Feng, C. Zhao, and H. Ding, "High-Speed Electromagnetic Transient (EMT) Equivalent Modelling of Power Electronic Transformers," *IEEE Transactions on Power Delivery,* vol. 36, no. 2, pp. 975–986, 2021.

90. G. Daoud, E. H. Aboadla, S. Khan, S. F. Ahmed, and M. Tohtayong, "Input-series output-parallel full-bridge DC-DC converter for high power applications," *2017 4th IEEE International Conference on Engineering Technologies and Applied Sciences (ICETAS),* 2017, pp. 1–6.

91. C. Gammeter, F. Krismer, and J. W. Kolar, "Comprehensive Conceptualization, Design, and Experimental Verification of a Weight-Optimized All-SiC 2 kV/700 V DAB for an Airborne Wind Turbine," *IEEE Journal of Emerging and Selected Topics in Power Electronics,* vol. 4, no. 2, pp. 638–656, 2016.

92. W. Chen, X. Ruan, H. Yan, and C. K. Tse, "DC/DC Conversion Systems Consisting of Multiple Converter Modules: Stability, Control, and Experimental Verifications," *IEEE Transactions on Power Electronics,* vol. 24, no. 6, pp. 1463–1474, 2009.

93. R. Giri, V. Choudhary, R. Ayyanar, and N. Mohan, "Common-duty-ratio control of input-series connected modular DC-DC converters with active input voltage and load-current sharing," *IEEE Transactions on Industry Applications,* vol. 42, no. 4, pp. 1101–1111, 2006.

94. D. Ochoa, A. Barrado, A. Lázaro, R. Vázquez, and M. Sanz, "Modeling, Control and Analysis of Input-Series-Output-Parallel-Output-Series architecture with Common-Duty-Ratio and Input Filter," *2018 IEEE 19th Workshop on Control and Modeling for Power Electronics (COMPEL),* 2018, pp. 1–6.

95. J. Shi, J. Luo, and X. He, "Common-Duty-Ratio Control of Input-Series Output-Parallel Connected Phase-shift Full-Bridge DC–DC Converter Modules," *IEEE Transactions on Power Electronics,* vol. 26, no. 11, pp. 3318–3329, 2011.

96. J. W. Kimball, J. T. Mossoba, and P. T. Krein, "A Stabilizing, High-Performance Controller for Input Series-Output Parallel Converters," *IEEE Transactions on Power Electronics,* vol. 23, no. 3, pp. 1416–1427, 2008.

97. M. Abrehdari, and M. Sarvi, "Comprehensive sharing control strategy for input-series output-parallel connected modular DC–DC converters," *IET Power Electronics,* vol. 12, no. 12, pp. 3105–3117, 2019.

98. W. Chen, X. Zhu, G. Wang, W. Jiang, and K. Yao, "A novel input voltage sharing control strategy for input-series output-parallel system with high reliability," *2014 IEEE Energy Conversion Congress and Exposition (ECCE),* 2014, pp. 1207–1212.

99. R. Ding, F. Wang, N. Zhang, S. Shi, S. Cheng, and F. Zhuo, "A Decentralized Control Strategy with Output Voltage Deviation-Correction for Input-Series-Output-Parallel DC Transformer Based on Dual-Active-Bridge," *2020 IEEE 9th International Power Electronics and Motion Control Conference (IPEMC2020-ECCE Asia),* 2020, pp. 2458–2462.

100. L. Qu, D. Zhang, and Z. Bao, "Output Current-Differential Control Scheme for Input-Series–Output-Parallel-Connected Modular DC–DC Converters," *IEEE Transactions on Power Electronics,* vol. 32, no. 7, pp. 5699–5711, 2017.

101. P. J. Grbovic, "Master/Slave Control of Input-Series- and Output-Parallel-Connected Converters: Concept for Low-Cost High-Voltage Auxiliary Power Supplies," *IEEE Transactions on Power Electronics,* vol. 24, no. 2, pp. 316–328, 2009.

102. D. Sha, Z. Guo, and X. Liao, "Cross-Feedback Output-Current-Sharing Control for Input-Series-Output-Parallel Modular DC–DC Converters," *IEEE Transactions on Power Electronics,* vol. 25, no. 11, pp. 2762–2771, 2010.

103. S. H. Kim, B. J. Kim, and C. Y. Won, "A Study on Decentralized Inverse-Droop Control for Input Voltage Sharing of ISOP Converter in the Current Control Loop," *2019 10th International Conference on Power Electronics and ECCE Asia (ICPE 2019 - ECCE Asia),* 2019, pp. 2382–2387.

104. Y. Huang, C. K. Tse, and X. Ruan, "General Control Considerations for Input-Series Connected DC/DC Converters," *IEEE Transactions on Circuits and Systems I: Regular Papers,* vol. 56, no. 6, pp. 1286–1296, 2009.

105. L. Qu, and D. Zhang, "Input voltage sharing control scheme for input series and output series DC/DC converters using paralleled MOSFETs," *IET Power Electronics,* vol. 11, no. 2, pp. 382–390, 2018.

106. P. Zumel, L. Ortega, A. Lázaro, C. Fernández, A. Barrado, A. Rodríguez, and M. M. Hernando, "Modular Dual-Active Bridge Converter Architecture," *IEEE Transactions on Industry Applications,* vol. 52, no. 3, pp. 2444–2455, 2016.

107. C. Luo, and S. Huang, "Novel Voltage Balancing Control Strategy for Dual-Active-Bridge Input-Series-Output-Parallel DC-DC Converters," *IEEE Access,* vol. 8, pp. 103114–103123, 2020.

108. X. Zhu, H. Hu, H. Tao, and Z. He, "Stability Analysis of PV Plant-Tied MVdc Railway Electrification System," *IEEE Transactions on Transportation Electrification,* vol. 5, no. 1, pp. 311–323, 2019.

109. V. M. Iyer, S. Gulur, G. Gohil, and S. Bhattacharya, "An Approach Towards Extreme Fast Charging Station Power Delivery for Electric Vehicles with Partial Power Processing," *IEEE Transactions on Industrial Electronics,* vol. 67, no. 10, pp. 8076–8087, 2020.

110. A. Mäki, and S. Valkealahti, "Power Losses in Long String and Parallel-Connected Short Strings of Series-Connected Silicon-Based Photovoltaic Modules Due to Partial Shading Conditions," *IEEE Transactions on Energy Conversion,* vol. 27, no. 1, pp. 173–183, 2012.

111. P. Manganiello, M. Balato, and M. Vitelli, "A Survey on Mismatching and Aging of PV Modules: The Closed Loop," *IEEE Transactions on Industrial Electronics,* vol. 62, no. 11, pp. 7276–7286, 2015.

112. L. Gao, R. A. Dougal, S. Liu, and A. P. Iotova, "Parallel-Connected Solar PV System to Address Partial and Rapidly Fluctuating Shadow Conditions," *IEEE Transactions on Industrial Electronics,* vol. 56, no. 5, pp. 1548–1556, 2009.

113. M. O. Badawy, S. M. Bose, and Y. Sozer, "A Novel Differential Power Processing Architecture for a Partially Shaded PV String Using Distributed Control," *IEEE Transactions on Industry Applications,* vol. 57, no. 2, pp. 1725–1735, 2021.

114. Y. Jeon, and J. Park, "Unit-Minimum Least Power Point Tracking for the Optimization of Photovoltaic Differential Power Processing Systems," *IEEE Transactions on Power Electronics,* vol. 34, no. 1, pp. 311–324, 2019.

115. K. Sun, Z. Qiu, H. Wu, and Y. Xing, "Evaluation on High-Efficiency Thermoelectric Generation Systems Based on Differential Power Processing," *IEEE Transactions on Industrial Electronics,* vol. 65, no. 1, pp. 699–708, 2018.

116. E. Candan, P. S. Shenoy, and R. C. N. Pilawa-Podgurski, "A Series-Stacked Power Delivery Architecture With Isolated Differential Power Conversion for Data Centers," *IEEE Transactions on Power Electronics,* vol. 31, no. 5, pp. 3690–3703, 2016.

117. G. Chu, H. Wen, Y. Hu, L. Jiang, Y. Yang, and Y. Wang, "Low-Complexity Power Balancing Point-Based Optimization for Photovoltaic Differential Power Processing," *IEEE Transactions on Power Electronics,* vol. 35, no. 10, pp. 10306–10322, 2020.

118. C. Olalla, D. Clement, M. Rodriguez, and D. Maksimovic, "Architectures and Control of Submodule Integrated DC–DC Converters for Photovoltaic Applications," *IEEE Transactions on Power Electronics,* vol. 28, no. 6, pp. 2980–2997, 2013.

119. F. Krismer, and J. W. Kolar, "Closed Form Solution for Minimum Conduction Loss Modulation of DAB Converters," *IEEE Transactions on Power Electronics,* vol. 27, no. 1, pp. 174–188, 2012.

120. Q. Gu, L. Yuan, J. Nie, J. Sun, and Z. Zhao, "Current Stress Minimization of Dual-Active-Bridge DC–DC Converter Within the Whole Operating Range," *IEEE Journal of Emerging and Selected Topics in Power Electronics,* vol. 7, no. 1, pp. 129–142, 2019.

Unified FDDC Control Scheme for I²ACL Isolated DC-DC Converters

2

In this chapter, the existing I^2ACL isolated dc-dc converters are reviewed thoroughly, including the unidirectional I^2ACL converters and the bidirectional I^2ACL converters in Sect. 2.1. Moreover, in Sect. 2.2, the general current transferred features of these two-type converters are discussed, respectively. Since the average transferred current of these converters is just influenced by the middle ac inductance a little during transient process, the I^2ACL isolated dc-dc converter can be regarded as the first-order converter. Then, based on the discovered general characteristic, a unified FDDC control scheme is proposed for improving the dynamic performance of these I^2ACL isolated dc-dc converters. In addition, the specialized design principles of the PI parameters in the unified FDDC control method are presented. Finally, to verify the universality and feasibility of the proposed general FDDC control strategy, simulation or experiment results are presented with demonstration examples, e.g. full bridge type, DAB-type, and the three-phase DAB type dc-dc converters in Sect. 2.3. Then, the general logic structure of this Chapter can be summarized in Fig. 2.1.

2.1   The I²ACL Isolated DC-DC Converters

In terms of power transferring stages, the I^2ACL dc-dc converter can be divided into the dc-ac stage and the ac-dc stage. Generally, voltage-fed switching networks are required to obtain ac voltage from dc voltage, then the energy conversion between dc power and ac power can be realized. Thus, the potential building blocks of the I^2ACL dc-dc

© The Author(s), under exclusive license to Springer Nature Switzerland AG 2025   27
N. Hou, *High-Robust Control Schemes for Dual-Active-Bridge-Based DC–DC Converter Systems in Renewable Energy Applications*, Synthesis Lectures on Power Electronics,
https://doi.org/10.1007/978-3-031-72963-8_2

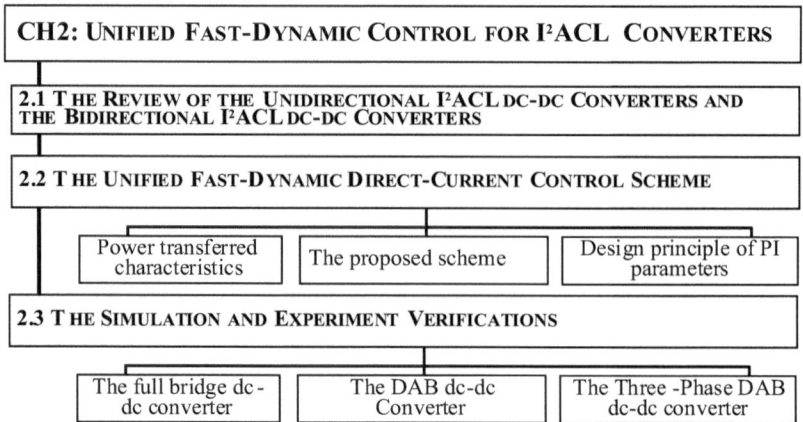

Fig. 2.1 Logic structure of this chapter

converter are analyzed first in this section. Moreover, based on these blocks, the existing unidirectional I²ACL dc-dc converters and bidirectional I²ACL dc-dc converters are reviewed.

2.1.1 The Potential Half Bridges for the I²ACL Isolated DC-DC Converter

The existing half bridges which can be employed to build the I²ACL isolated dc-dc Converters are demonstrated in Fig. 2.2. Five types can be considered, including the diode-based half bridge, hybrid half bridge, switch-based half bridge, neutral point clamped (NPC) half bridge [1], and T-type half bridge [2]. (It is noted for interested readers that other switching networks can also be employed to obtain ac voltage from dc voltage, such as multilevel NPC bridge [3] and active NPC bridge [4], etc.) As shown in Fig. 2.2, the first three half bridges can usually achieve two voltage levels as 0 and U_{dc}, and the latter two half bridges can usually achieve three voltage levels as 0, $U_{dc}/2$, and U_{dc}. Then, based on these half bridges, different switching networks can be obtained for forming ac voltage.

Basically, combing the series capacitors and the half bridge, the simplest bridges that can be obtained for acquiring the ac output voltage is shown in Fig. 2.3. The first three bridges can obtain the ac voltage by $-U_{dc}/2$ and $U_{dc}/2$, and the latter two bridges can generate the ac voltage by $-U_{dc}/2$, 0, and $U_{dc}/2$. Since these bridges are constructed by one half bridge in Fig. 2.22, the I²ACL dc-dc converter constructed by these bridges can be called half-bridge dc-dc converter.

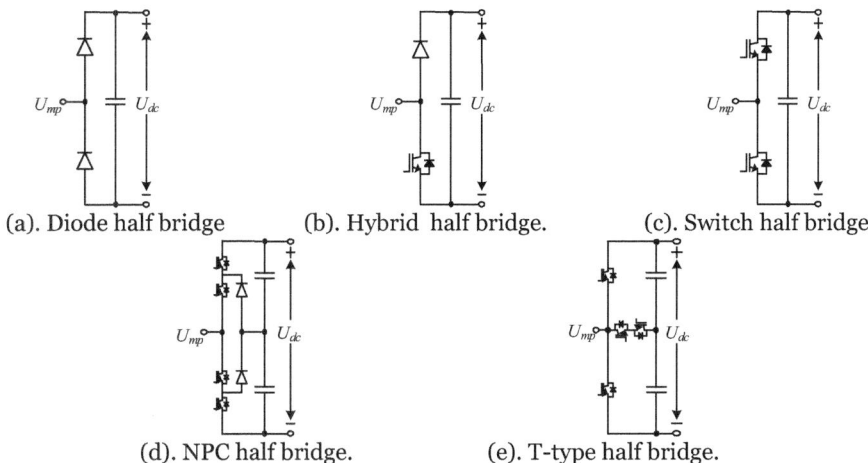

Fig. 2.2 The basic half bridges for establishing the switching network for I²ACL dc-dc converter

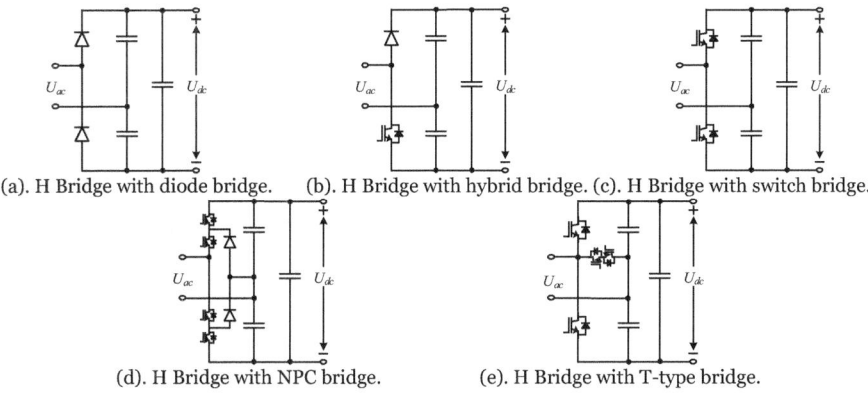

Fig. 2.3 The switching network with one half bridge for forming ac voltage

Moreover, based on any two half bridges in Fig. 2.2, the full bridges can be acquired. Besides, the full bridges with the same half bridge are shown in Fig. 2.4, which are the most used full bridges for establishing the existing isolated dc-dc converters. Compared with the first three H bridges, the latter two H bridges including the NPC full bridges and the T-Type full bridges can generate multi-level ac voltages, and the I²ACL dc-dc converters established by the latter two full bridges are usually called multi-level dc-dc converters. Some I²ACL isolated dc-dc converters are established by the full bridges with different half bridges, and these full bridges will be mentioned in the review of the specific I²ACL dc-dc converters. Similarly, with multiple half bridges, the multi-phase I²ACL dc-dc converter can be acquired.

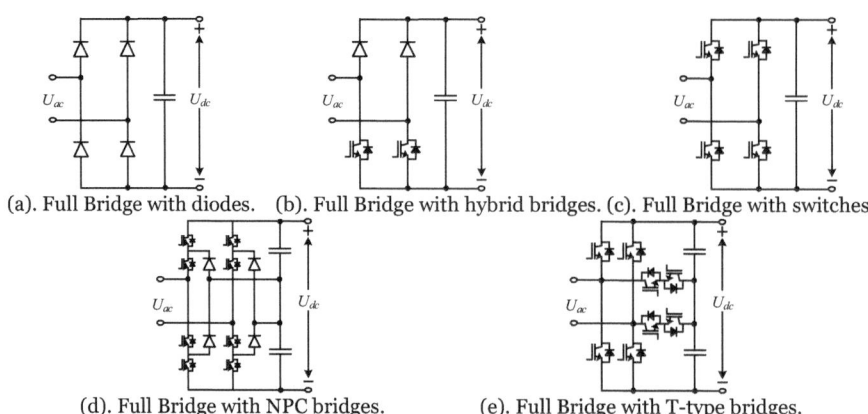

(a). Full Bridge with diodes.   (b). Full Bridge with hybrid bridges. (c). Full Bridge with switches.

(d). Full Bridge with NPC bridges.   (e). Full Bridge with T-type bridges.

Fig. 2.4 The switching network with two half bridges for I²ACL dc-dc converter

In addition, when only diodes are employed to establish the H bridges as shown in Fig. 2.2a, zero level voltage cannot be provided, which usually limits the voltage range of I²ACL dc-dc converters with these diode-based H bridges. Thus, diode-based H bridges with extra switches are required to boost the output range of these I²ACL dc-dc converters as shown in Fig. 2.5.

Notably, although the half bridges (shown in Figs. 2.3, 2.4 and 2.5) usually acquire voltage from the same dc bus or capacitor, these half bridges in the same H bridges

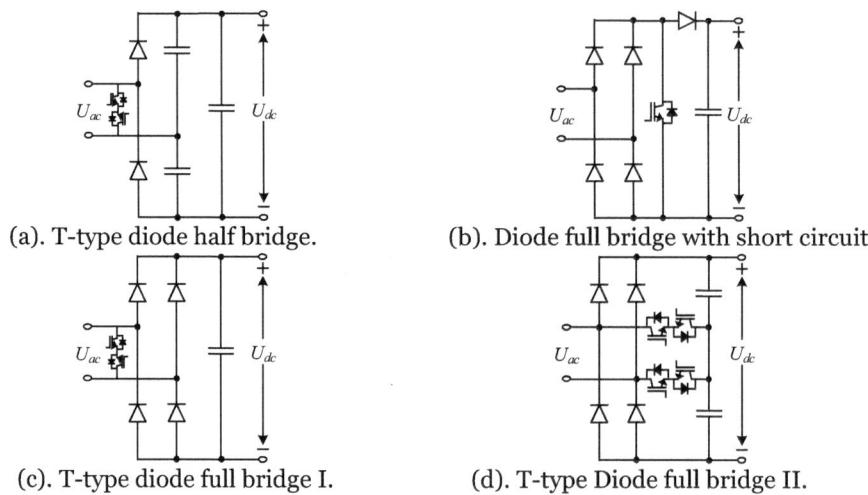

(a). T-type diode half bridge.   (b). Diode full bridge with short circuit.

(c). T-type diode full bridge I.   (d). T-type Diode full bridge II.

Fig. 2.5 The switching network based on diode half Bridges embedded with zero-level voltage

Fig. 2.6 The full bridge with
different dc-link voltages

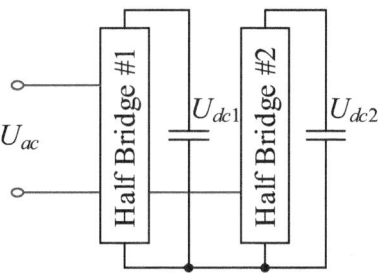

can connect with different dc buses as shown in Fig. 2.6, which may be suitable for
applications with multi-port sources.

2.1.2   The Unidirectional I²ACL Isolated DC-DC Converters

In this section, the existing unidirectional I²ACL dc-dc converters are reviewed, and the
simplified circuit of one branch of these converters can be expressed as shown in Fig. 2.7,
where the primary side of the unidirectional I²ACL dc-dc converter is established by the
controllable switches as shown in Fig. 2.2c–e, and the secondary side of the unidirectional I²ACL dc-dc converter is constructed by the uncontrollable diodes as shown in
Fig. 2.2a–b. Since diodes can only deal with the unidirectional power flow, the power of
the unidirectional I²ACL dc-dc converter can only be transferred from the primary side to
the secondary side. Then, the unidirectional I²ACL dc-dc converter is usually employed
to connect the renewable sources and the dc links such as photovoltaic system or fuel cell
system for maximum power point tracking performance [5, 6], and when the unidirectional I²ACL dc-dc converter is connected to the power consumer side, this converter can
also be used to adjust the dc-link voltage [7, 8].

Simply, combining the switching networks as shown in Fig. 2.3a and c, the asymmetric
half bridge dc-dc converter with diode half bridge can be obtained as shown in Fig. 2.8
[9]. Further, by switching the secondary-side H Bridge to the diode full bridge as shown
in Fig. 2.4a, the asymmetric half bridge dc-dc converter with diode full bridge can be
acquired as shown in Fig. 2.9 [9]. Similarly, by switching the primary-side H Bridge to
the switch full bridge as shown in Fig. 2.4c, the full bridge dc-dc converter with diode half

Fig. 2.7 The simplified circuit
of unidirectional I²ACL dc-dc
converters

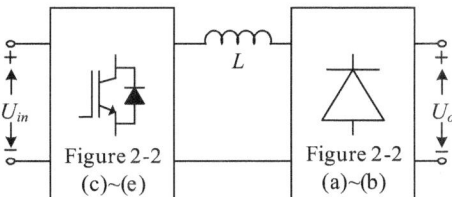

bridge can be shown in Fig. 2.10 [10]. Besides, combining the switch full bridge and the diode full bridge, the full bridge dc-dc converter with diode full bridge can be generated as shown in Fig. 2.11 [11], which may be the most popular unidirectional isolated dc-dc converter in industrial applications.

Moreover, when the primary side and the secondary side both have three half bridges, the three-phase unidirectional dc-dc converter with a three-phase diode bridge can be obtained as shown in Fig. 2.12a. By switching the primary-side three-phase bridge to three dual active bridges, another three-phase unidirectional dc-dc converter can be obtained as shown in Fig. 2.12b [12].

Since the secondary-side bridges of the topologies as shown in Figs. 2.8, 2.9, 2.10, 2.11 and 2.12 are only established by the diodes, these bridges cannot provide zero voltage, which usually limits the output-voltage range of these converters. Therefore, some

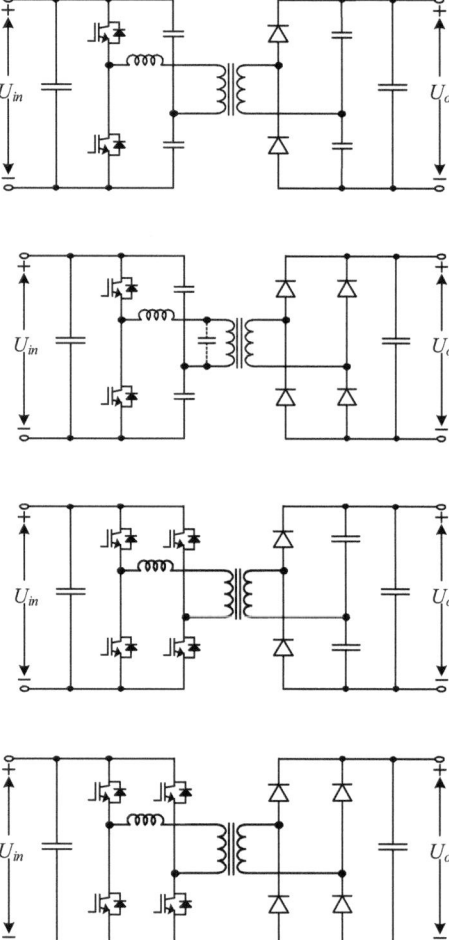

Fig. 2.8 The topology of the asymmetric half-bridge dc-dc converter with diode half bridge

Fig. 2.9 The topology of the asymmetric half-bridge dc-dc converter with diode full bridge

Fig. 2.10 The topology of the full bridge dc-dc converter with diode half bridge

Fig. 2.11 The topology of the full bridge dc-dc converter with diode full bridge

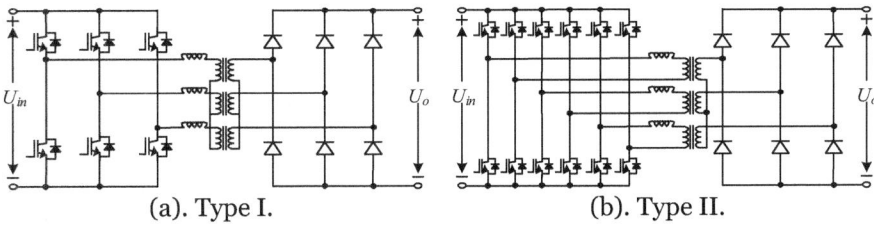

Fig. 2.12 The topology of the three-phase dc-dc converters with diode three-phase bridge

extra switches can be employed to boost the output range of these converters, and the potential bridges can be selected from Figs. 2.4 and 2.5. Moreover, the full-bridge dc-dc converters with active boost rectifier are shown in Fig. 2.13 [13, 14]. Similarly, as shown in Fig. 2.14, a three-phase semi-dual active bridge dc-dc converter is presented for wide input variations and high voltage interface [15].

In addition, as shown in Fig. 2.6, the half bridges in the same switching network can be connected to different dc links, which can be employed in multiple-port converter

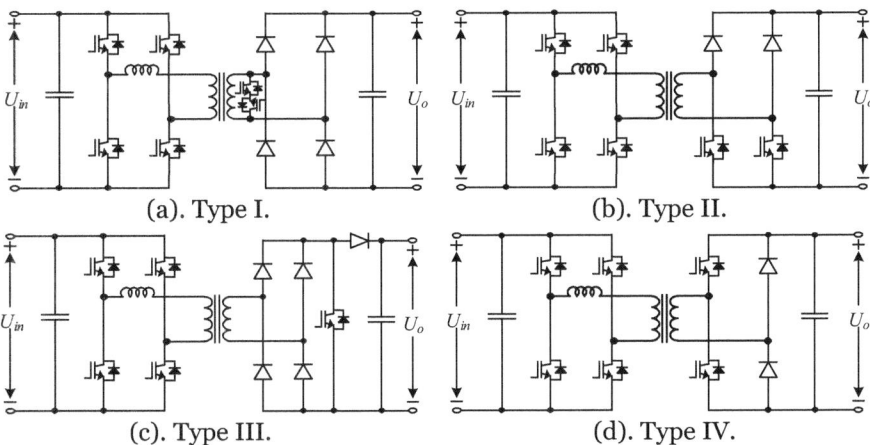

Fig. 2.13 The topologies of single-phase unidirectional I²ACL dc-dc converters with active boost rectifier

Fig. 2.14 The topologies of three-phase unidirectional I²ACL dc-dc converters with active boost rectifier

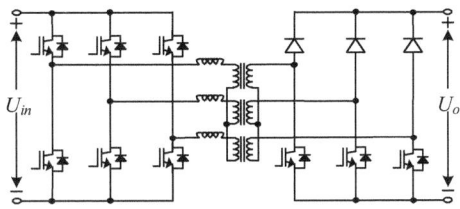

system and can be used to generate multiple-level voltages. Moreover, a secondary-side modulated full-bridge dc-dc converter is shown in Fig. 2.15a [16]. However, this full-bridge dc-dc converter cannot generate symmetrical three-level voltages on the secondary side, the modified secondary-side modulated full-bridge dc-dc converters can be shown in Fig. 2.15b [17].

With electrical isolation, the unidirectional I²ACL dc-dc converter can be easily designed as a module with parallel or series configurations for high-power and high-voltage applications, especially one-side parallel and one-side series [18]. There are also some variant unidirectional I²ACL dc-dc converters with multiple H bridges for high-power high-voltage applications. An interleaved full-bridge converter with diode half bridges can be shown in Fig. 2.16 [19], and a full-bridge dc-dc converter with paralleled input IGBTs and split secondary windings can be shown in Fig. 2.17 [20].

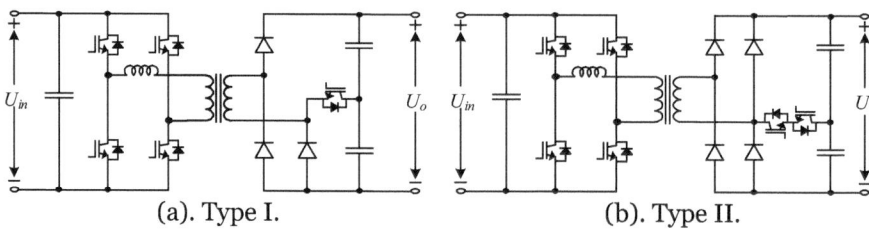

(a). Type I.   (b). Type II.

Fig. 2.15 The topology of the secondary-side modulated full bridge dc-dc converter

Fig. 2.16 Interleaved
full-bridge converter with
diode half bridges

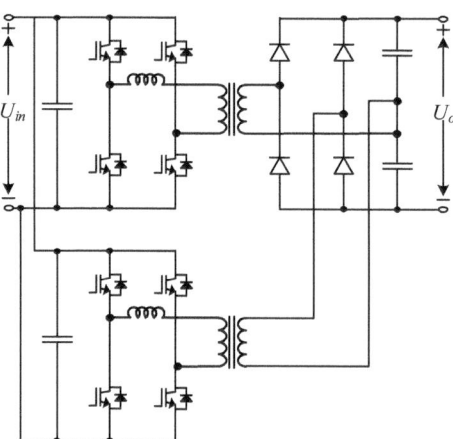

Fig. 2.17 Full-bridge
converter with paralleled input
IGBTs and split secondary
windings

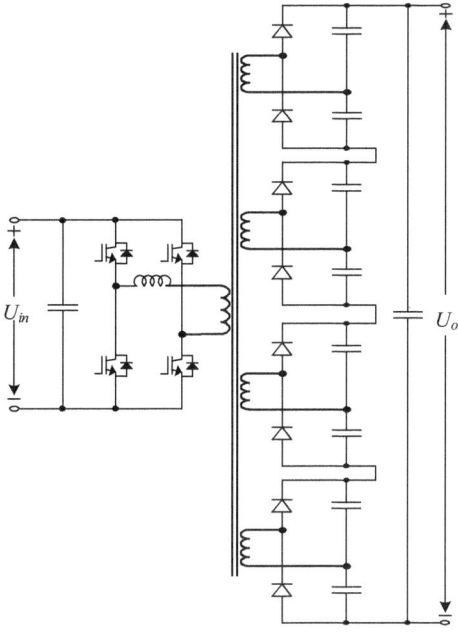

2.1.3   The Bidirectional I²ACL Isolated DC-DC Converters

In this section, the existing bidirectional I²ACL dc-dc converters are reviewed, and the simplified circuit of one branch of this kind of converter can be expressed as Fig. 2.18, where both the primary side and the secondary side of these converters are constructed with controllable half bridges as shown in Fig. 2.2c–e. Based on two half switch bridges, the symmetrical half-bridge dc-dc converter can be shown in Fig. 2.19 [21]. Then, by changing the primary-side H Bridge as the dual active bridge, the unsymmetrical dual active bridge dc-dc converter can be shown in Fig. 2.20 [22]. Moreover, when both sides are dual active bridges, the dual active bridge dc-dc converter can be acquired as shown in Fig. 2.21 [23], which is regarded as one of the most promising dc-dc converters.

Similarly, when the primary side and the secondary side both have three half bridges, the well-known three-phase dual-active bridge dc-dc converter can be shown in Fig. 2.22a

Fig. 2.18   The simplified
circuit of bidirectional I²ACL
dc-dc converters

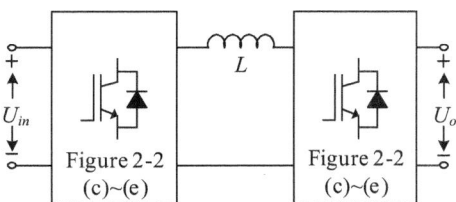

Fig. 2.19 The topology of the
symmetrical half-bridge dc-dc
converter

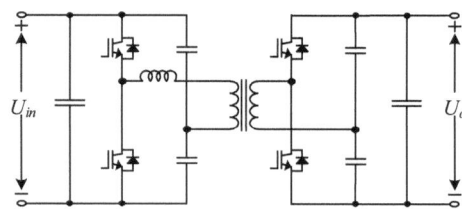

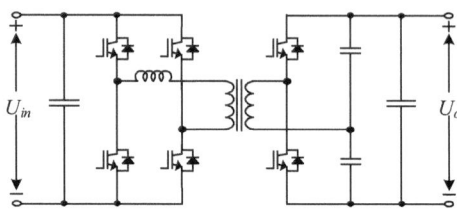

Fig. 2.20 The topology of the unsymmetrical dual-active-bridge dc-dc converter with switch half
bridge

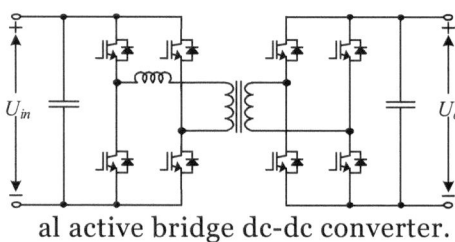

al active bridge dc-dc converter.

Fig. 2.21 The topology of the dual active bridge dc-dc converter

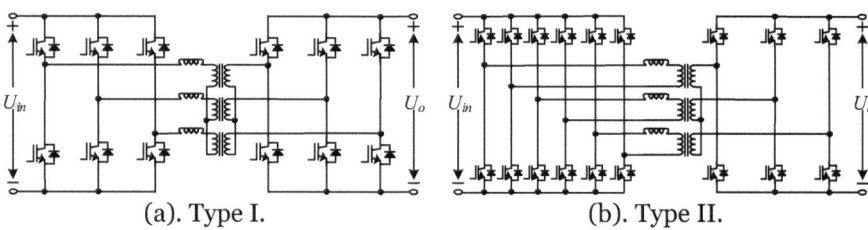

(a). Type I.   (b). Type II.

Fig. 2.22 The topology of the three-phase dual active bridge dc-dc converters

[24]. Then, by switching the primary-side three-phase bridge as three dual active bridges,
another three-phase bidirectional dc-dc converter can be obtained as shown in Fig. 2.22b
[25].

Moreover, based on T-type half bridge and NPC half bridge as shown in Fig. 2.2d and e, some multilevel isolated dc-dc converters can be obtained as shown in Figs. 2.23, 2.24, 2.25 and 2.26. Combining two T-type half bridge, a three-level symmetrical T-type isolated dc-dc converter can be shown in Fig. 2.23 [26], which has a smaller number of switches. Besides, by combing the NPC half bridge and the full bridge, a three-level unsymmetrical NPC dc-dc converter can be shown in Fig. 2.24 [27], which can be employed to connect the low voltage bus and the high voltage bus. Moreover, combining the full bridge and the full NPC bridge, the five-level unsymmetrical NPC dc-dc converter can be obtained as shown in Fig. 2.25 [28]. By switching the full NPC bridge to the T-type bridge, the five-level unsymmetrical T-type DAB dc-dc converter can be obtained as shown in Fig. 2.26 [29]. In addition, combining two full NPC bridges, the five-level symmetrical NPC dc-dc converter can be obtained as shown in Fig. 2.27 [30], which is a promising candidate for the high voltage dc-dc applications. Similarly, the five-level symmetrical T-type dc-dc converter can be obtained as shown in Fig. 2.28.

In addition, based on multiple-winding transformers and different half bridges, some I²ACL dc-dc converters can be obtained as shown in Figs. 2.29, 2.30 and 2.31, which is usually employed to reduce the current stress of the switches or increase the power density of the converter system. As shown in Fig. 2.29, a three-winding-transformer-based dual

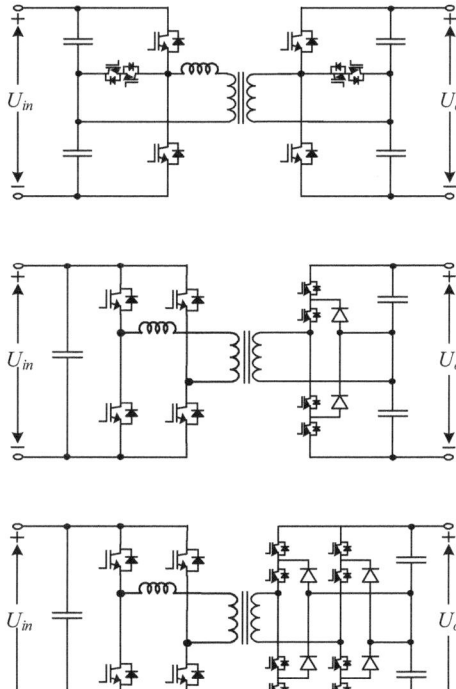

Fig. 2.23 The topology of the three-level symmetrical T-type dc-dc converter

Fig. 2.24 The topology of the three-level unsymmetrical NPC dc-dc converter

Fig. 2.25 The topology of the five-level unsymmetrical NPC dc-dc converter

Fig. 2.26 The topology of the
five-level unsymmetrical
T-type dual active bridge dc-dc
converter

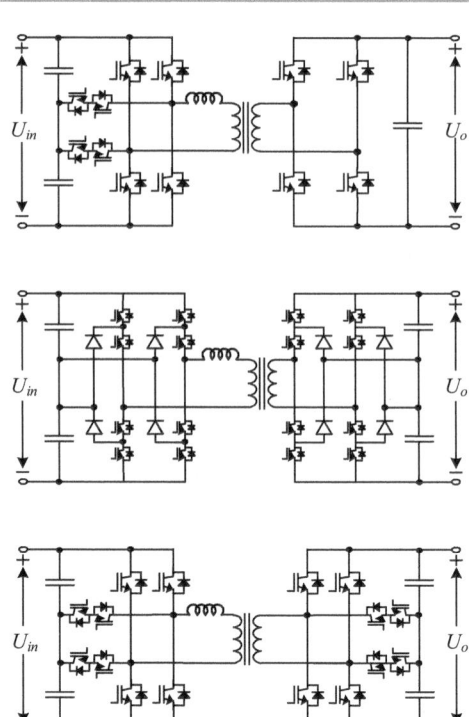

Fig. 2.27 The topology of the
five-level symmetrical NPC
dc-dc converter

Fig. 2.28 The topology of the
five-level symmetrical T-type
dc-dc Converter

active bridge dc-dc converter is presented for reducing the current stress of switches on
the secondary side [31]. Moreover, a multi-winding-transformer-based dual active bridge
dc-dc converter with paralleled output configuration can be shown in Fig. 2.30, which
can be employed to reduce the current stress at the low-voltage side when the difference
between the input voltage and the output voltage is very large [32]. Then, based on the
three-winding transformer, a three-port dual active bridge dc-dc converter can be obtained
for arranging the power transmission among three independent dc terminals as shown in
Fig. 2.31 [33].

As shown in Fig. 2.6, the half bridges in the same switching network can be connected to different dc links, which can be employed in the multi-port converter system

Fig. 2.29 The three-winding-
transformer-based dual active
bridge dc-dc converter

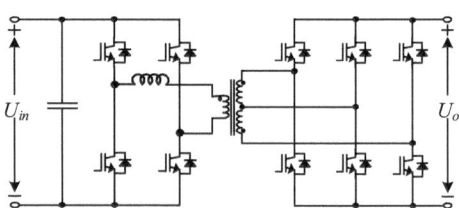

Fig. 2.30 The multi-winding-transformer-based dual active bridge dc-dc converter with paralleled output configuration

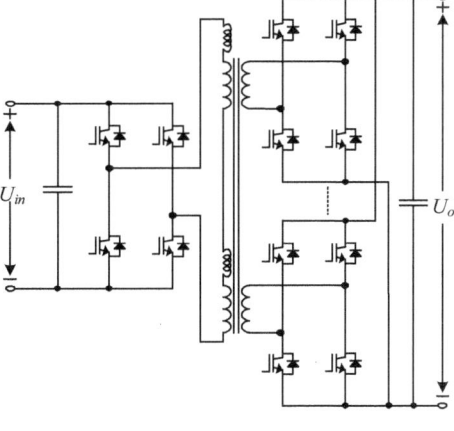

Fig. 2.31 The three-port dual active bridge dc-dc converter

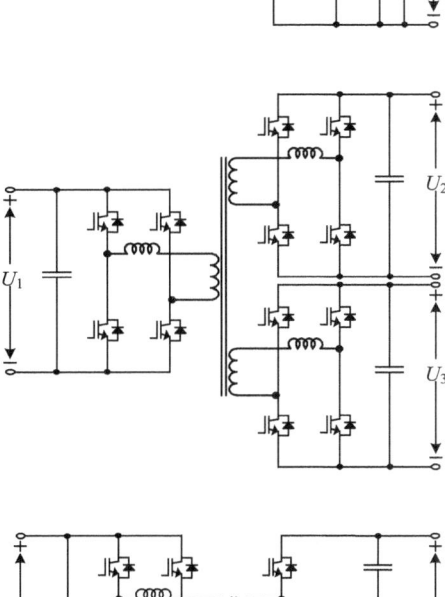

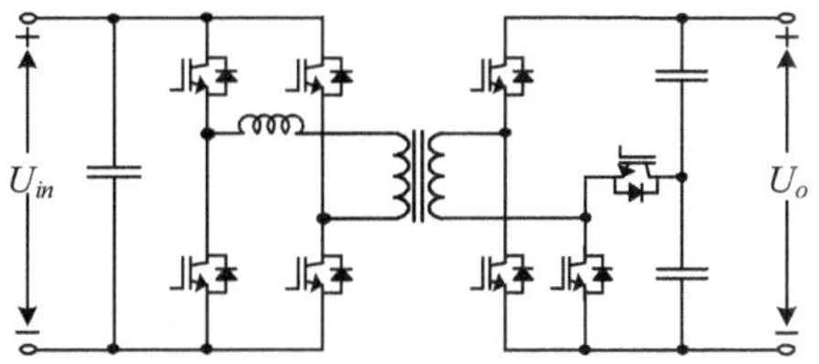

Fig. 2.32 The topology of the secondary-side modulated bidirectional full-bridge dc-dc converter

and can be used to generate multiple-level voltages. Similarly, a secondary-side modulated bidirectional full-bridge dc-dc converter can be shown in Fig. 2.32. [16].

2.1.4 The Summary of the Unidirectional and Bidirectional I²ACL Isolated DC-DC Converters

Based on the previous overview, these existing unidirectional and bidirectional I²ACL isolated dc-dc converters can be summarized in Fig. 2.33. Based on the voltage level, the phase number, and the winding number, these existing topologies are divided into five categories including two-level I²ACL converter, three-level I²ACL converter, multi-level I²ACL converter, multi-phase I²ACL converter, and multi-winding I²ACL converter.

Generally, the unidirectional I²ACL dc-dc converter can be employed in some unidirectional applications such as PV and fuel cell systems [9–17, 19, 20]. The bidirectional I²ACL dc-dc converter can be used in some bidirectional applications such as energy storage systems and dc grid systems [21–30, 32, 34, 35]. According to the voltage value, the two-level I²ACL isolated dc-dc converter and the three-level I²ACL isolated dc-dc converter can be employed in some low voltage conditions [11, 13, 21, 26]. Since the multi-level I²ACL isolated dc-dc converter can tolerate higher voltage, these converters can be suitable for some middle voltage applications [28, 29]. Moreover, compared with the single-phase I²ACL isolated dc-dc converter, the multi-phase I²ACL isolated dc-dc converter can provide a lower ripple current to the dc-link [15, 25], so a lower dc-link capacitor can usually be adopted. Sometimes, multi-ports may be required for connecting several voltage sources, where the multi-winding I²ACL isolated dc-dc converters can be used for high power density [20, 32, 34]. Compared with the single-phase two-winding I²ACL isolated dc-dc converters, the transformer design of the last two types will be usually more difficult.

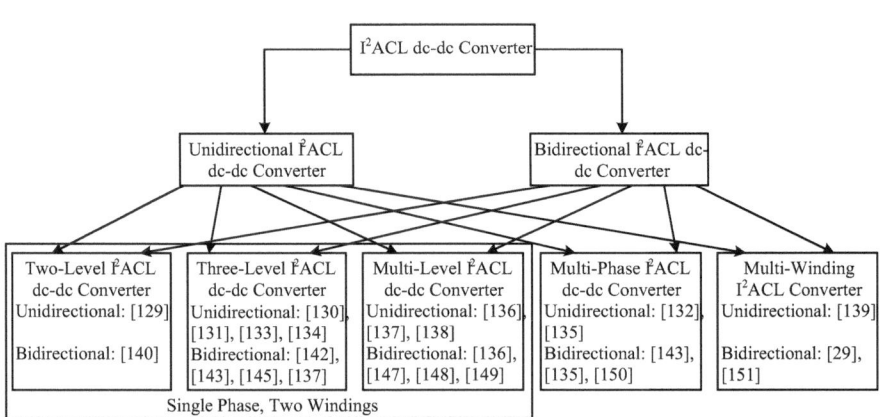

Fig. 2.33 Detailed classification of the existing I²ACL isolated dc-dc converters

2.2 The Unified FDDC Control Scheme

In this section, the unified FDDC control scheme is proposed for improving the dynamic response of the I^2ACL isolated dc-dc converter when the input voltage and the load condition are suddenly changed. As shown in Figs. 2.7 and 2.18, there is always a middle inductance in one power transferring branch of the I^2ACL dc-dc converters, so the influence on the transient process is analyzed first in this section. Besides, the order reducing phenomenon of the I^2ACL dc-dc converter can be obtained since the middle inductance can only influence the transient process a little. Moreover, based on this characteristic, a unified FDDC control scheme is proposed for all the I^2ACL dc-dc converters with the design principle of PI parameters. In addition, since the large efficiency difference usually influences the proposed general fast-dynamic control strategy, a compensation operation for efficiency difference is presented for ensuring the dynamic response of the proposed FDDC control scheme.

2.2.1 The Power Transferred Characteristics of I^2ACL Isolated DC-DC Converters

From the previous analysis, there are two kinds of I^2ACL dc-dc converters including the unidirectional I^2ACL dc-dc converter and the bidirectional I^2ACL dc-dc converter. Different from the unidirectional I^2ACL dc-dc converter, both sides of the bidirectional I^2ACL dc-dc converter can be positively controllable. So, in terms of the transient performance when the transferred power of the converter is suddenly changed for dealing with the disturbance of input voltage and load condition, there is a little difference between the unidirectional I^2ACL dc-dc converter and the bidirectional I^2ACL dc-dc converter. By using the full-bridge dc-dc converter as an example for the unidirectional I^2ACL dc-dc converter, when the transferred power is suddenly changed with the disturbance of input voltage and load condition, the transient waveforms of the full-bridge dc-dc converter can be shown in Fig. 2.34.

As shown in Fig. 2.34, when the required transferred power of the full bridge is suddenly changed, the required transferred power can be usually obtained in the second

Fig. 2.34 The transient waveforms when the transferred power of the full-bridge dc-dc converter is changed

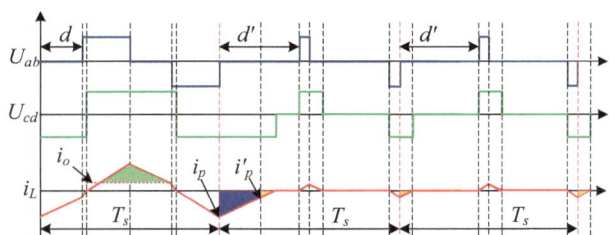

switching period when the phase-shift ratio d is changed. Moreover, in the first switching period, there is additional power stored in the middle inductance, which is transferred from the input side to the output side. The additional power P_a can be calculated as,

$$P_a = \frac{L(i_p^2 - i_p'^2)}{2T_s} \tag{2.1}$$

where i_p and i'_p are the peak currents of the middle inductance before and after the disturbance of phase-shift ratio d, respectively. The output-voltage disturbance ΔU_o caused by the additional power can be expressed as,

$$\Delta U_o \approx \frac{L(i_p^2 - i_p'^2)}{2C_o U_o} \tag{2.2}$$

As shown in Fig. 2.34, the green part of the middle-inductance current usually results in the output-voltage ripple. So, output-voltage disturbance ΔU_o caused by the additional power from the middle inductance is usually similar to the output-voltage ripple, and this additional power cannot influence the transient performance of the full-bridge dc-dc converter.

Moreover, by using the DAB dc-dc converter as an example for bidirectional I²ACL dc-dc converter, when the transferred power suddenly changes for addressing the disturbance of input voltage and load condition, the transient waveforms of the DAB dc-dc converter can be shown in Fig. 2.35.

As shown in Fig. 2.35, since the input voltage and the output voltage can be regarded as the same in a switching period, the transferred power P_T of the DAB dc-dc converter can be calculated by the new phase-shift ratio D' as,

$$P_T = \frac{1}{T_s} \int_0^{T_s} \frac{U_{cd}}{n}(i_L + \Delta i_L)dt = \frac{1}{T_s} \int_0^{T_s} U_{ab}(i_L + \Delta i_L)dt$$

$$= \frac{2}{T_s} \int_0^{\frac{T_s}{2}} U_{in}i_L dt + \frac{\Delta i_L}{T_s} \int_0^{T_s} U_{ab}\,dt$$

Fig. 2.35 The transient waveforms when the transferred power of the DAB dc-dc converter is changed

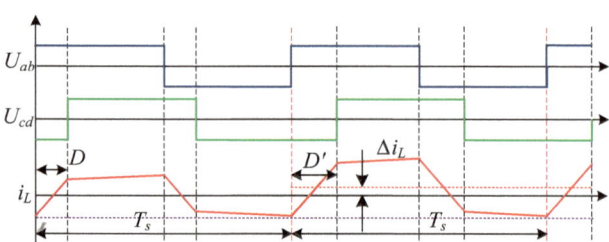

Fig. 2.36 The equivalent circuit for the dc offset of the inductance current at the steady-state condition

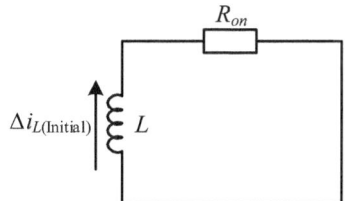

$$= \frac{2}{T_s} \int_0^{\frac{T_s}{2}} U_{in} i_L dt = \frac{U_{in} U_o D(1-D)T_s}{2nL} \quad (2.3)$$

According to (2.3), since the dc offset current of the middle inductance cannot influence the power transmission of the DAB dc-dc converter, the transferred power of the DAB dc-dc converter can be controlled by the phase-shift ratio accurately at the steady-state condition and during the transient process [36]. In addition, the dc offset of the inductance current can be consumed by the conducting resistor R_{on} of the DAB dc-dc converter, and since U_{ab} and U_{cd} are total ac components at steady state condition which cannot generate dc inductance current, the equivalent circuit can be shown as Fig. 2.36. Then, the dc offset of the inductance current Δi_L can be consumed by the conducting resistor R_{on} of the DAB dc-dc converter gradually, which affects the transferred power of this converter slightly with the tiny conducting resistor.

Therefore, for the I^2ACL dc-dc converter including the unidirectional I^2ACL dc-dc converter and the bidirectional I^2ACL dc-dc converter, when the transferred power is suddenly changed for dealing with the variation of input voltage and load condition, the required transferred power can be obtained within two switching periods. The required transferred current of the I^2ACL dc-dc converter can be calculated as,

$$i_T = \frac{P_T}{U_o} \quad (2.4)$$

In (2.4), since output voltage can remain at its desired value with the sudden change of suitable transferred power, the required transferred current can also be obtained within two switching periods for a certain phase-shift ratio. Thus, the transferred current i_T of the I^2ACL converter can be directly controlled by the phase-shift value timely. With current-level modulation, the I^2ACL dc-dc converter can be regarded as the controllable current source [37], and the simplified circuit of the I^2ACL isolated dc-dc converter can be demonstrated in Fig. 2.37. So, although there is the middle inductance in the I^2ACL isolated dc-dc converter, the middle inductance doesn't influence the transient performance of this kind of converter, which is very different from the traditional dc-dc converters such as BUCK and BOOST. Based on this characteristic, the order reducing phenomenon can be obtained, and the fast-dynamic performance can be easily provided for the I^2ACL dc-dc converter.

Fig. 2.37 The simplified circuit of the I²ACL dc-dc converter

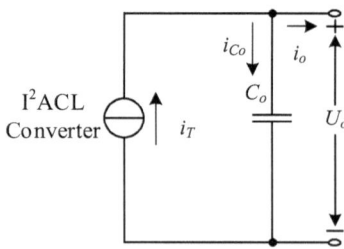

2.2.2   The Unified FDDC Control Method for I²ACL Isolated DC-DC Converter

Based on the power or current transferring characteristic of I²ACL dc-dc converters previously, a general FDDC control method is proposed for this type of converters, which can be employed to deal with the variations of the input voltage and the load condition. To face the load-condition change, the load current can be measured to calculate the desired output current i^*_o as,

$$i^*_o = \frac{U^*_o}{R} = \frac{i_o U^*_o}{U_o} \tag{2.5}$$

where R is the load equivalent resistor. Based on the Law of Conservation of Energy, the required transferred current of the I²ACL dc-dc converter should be the same as the desired output current. However, there are always some power losses in the converter system, which results in a little difference between the transferred current and the output current. So, the voltage deviation is unavoidable. To compensate for the error caused by the power loss and other uncertain values, the PI controller is used for obtaining a compensation coefficient k_{io} for acquiring the actual required transferred current. Moreover, the diagram for obtaining the actual required transferred current i^*_T can be shown in Fig. 2.38. Moreover, the phase-shift modulation method is the most popular modulation method for the I²ACL dc-dc converter, and based on the phase-shift ratio, the transferred current of these dc-dc converters can be directly obtained [38]. In reverse, when the required transferred current is obtained, the phase-shift ratio can be calculated. In addition, the complete diagram of the unified FDDC control method for the I²ACL dc-dc converter can be shown in Fig. 2.39.

As shown in Fig. 2.39, the general FDDC control scheme can be realized for the I²ACL dc-dc converter. At the beginning of each switching period, the input voltage U_{in}, the output voltage U_o and the load current i_o are measured. Based on the PI controller, the compensation factor k_{io} can be obtained by the output voltage U_o and its desired value U^*_o. Then, according to (2.5), the required load current i^*_o can be calculated by the actual load current i_o, the output voltage U_o and its desired value U^*_o. Moreover,

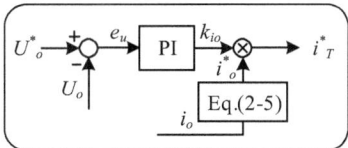

Fig. 2.38 The diagram for obtaining the actual required transferred current i^*_T

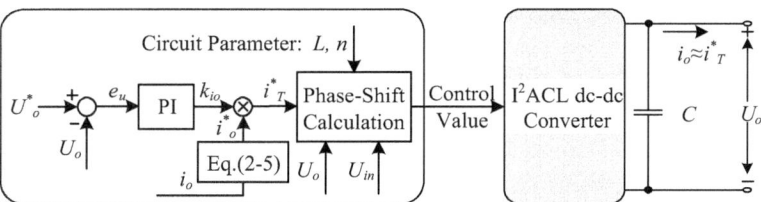

Fig. 2.39 The diagrams of the unified FDDC control schemes for the I^2ACL dc-dc converter

by combining the compensation factor k_{io} and the required load current i^*_o, the required transferred current i^*_T of the I^2ACL dc-dc converter can be obtained. In addition, based on the relationship between the transferred current i_T and the phase-shift ratio of the I^2ACL dc-dc converter, the required phase-shift ratio can usually be obtained by using the circuit parameter, the output voltage U_o, the input voltage U_{in} and the required transferred current i^*_T. Finally, based on the phase-shift modulation part, the required transferred current i^*_T can be realized, which can meet the requirement of the load consumption. Thus, the ultrafast dynamic response can be provided for the I^2ACL dc-dc converter when the input voltage and load condition are changed. Besides, since the compensation part is multiplied with the feed-forward load current, this unified FDDC control scheme is not sensitive to the circuit parameter [8, 36].

2.2.3 The Design Principle of PI Parameters

As shown in Fig. 2.39, when the transferred power of the I^2ACL dc-dc converter is controlled by the phase-shift ratio accurately and timely, the middle inductance can be omitted in its simplified circuit. Then, only the output capacitor can influence the dynamic performance of the I^2ACL dc-dc converter, and the dynamic model of the output capacitor can be expressed as,

$$C_o \frac{dU_o}{dt} = i_T - i_o \tag{2.6}$$

Based on the proposed general fast-dynamic control scheme, the transferred current of the I²ACL dc-dc converter can meet the requirement of the load condition immediately, and according to (2.6), the disturbance of output voltage during the transient process is limited. Therefore, the PI parameters in the general fast-dynamic control scheme cannot be determined based on the transient process, which is different from the traditional way of designing the PI parameter.

Moreover, when the measurement noise is considered, there should be irregular oscillations in phase-shift ratio at steady-state conditions, which may result in the irregular oscillations of output voltage. With the filter function of the output capacitor, the oscillation of the output voltage can be avoided, but irregular oscillations in the phase-shift ratio are inevitable. Thus, the disturbances of the phase-shift ratio ΔPSR caused by the measurement noises should be treated as a criterion to evaluate the stability of the I²ACL dc-dc converter. Then, the disturbance of required transferred current Δi^*_T caused by the measurement noise of output voltage U_{om} can be expressed as,

$$\Delta i^*_T = U_{om}(k_i + k_p)i^*_o \tag{2.7}$$

where k_i is the integral parameter and k_p is the proportional parameter of the PI controller. Based on the relationship between the transferred power and the phase-shift ratio of I²ACL dc-dc converter, the disturbance of phase-shift ratio ΔPSR can be obtained by using the required transferred current i^*_T and its disturbance Δi^*_T as,

$$\Delta PSR = f(\Delta i^*_T, i^*_T) \le \Delta PSR_{limt} \tag{2.8}$$

where ΔPSR$_{limt}$ is the limited value of the disturbance of the phase-shift ratio caused by the measurement noise of the output voltage. Combining (2.7) and (2.8), k_i and k_p should meet the requirement as,

$$k_i + k_p \le \frac{U_{om}i^*_o}{f^{-1}(\Delta PSR_{limt}, i^*_T)} \tag{2.9}$$

According to (2.9), the upper limitation of k_i and k_p can be obtained, and then, the disturbance of the phases-shift ratio can be restricted.

2.2.4 The Compensation Operation for the Efficiency Difference Caused by the Power Loss

In the proposed general fast-dynamic control scheme, the PI controller is employed to compensate the difference between the transferred current i_T and the output current i_o caused by the power loss, and the efficiency η of the I²ACL dc-dc converter can be approximatively expressed as,

$$\eta \approx \frac{P_o}{P_T} = \frac{U_o i_o}{U_o i_T} = \frac{1}{k_{io}} \tag{2.10}$$

According to (2.10), when the efficiency η of the new steady-state condition is a little different from its previous value, the new compensation coefficient k_{io} should also be a little different from its previous value. Besides, when the input voltage and load condition are changed, more time is needed for obtaining the required coefficient k_{io} based on the PI controller. Therefore, to reduce the settling time under the general fast-dynamic control scheme, a general compensation method is proposed for the I^2ACL dc-dc converter. As shown in Fig. 2.37, the relationship among the transferred current i_T, the capacitor charging current $i_{Co,}$ and the load current i_o can always be expressed as,

$$i_{Co} = i_T - i_o \tag{2.11}$$

Moreover, with the output capacitor C_o, the output voltage of the I^2ACL dc-dc converter can be kept at its desired voltage in the previous switching periods after the variation of the input voltage or the load resistor. Therefore, when the capacitor charging current i_{Co} is not equivalent to zero, the transferred current i_T of the I^2ACL dc-dc converter should compensate for this capacitor current i_{Co}. Thus, the transferred current i_T can be expressed as,

$$i_T = k_{io} i_o - i_{Co} = k_{io} i_o - \frac{C_o}{T_s}(U_o - U_o') \tag{2.12}$$

here U'_o is the measured output voltage in the last switching period. The required compensation coefficient can be expressed as,

$$k'_{io} = \frac{1}{i_o}\left[k_{io} i_o - \frac{C_o}{T_s}(U_o - U_o')\right] \tag{2.13}$$

When the load current i_o or the input voltage U_{in} is changed, (2.13) can be used to calculate the required compensation coefficient k_{io} for dealing with the efficiency difference under different conditions for the I^2ACL dc-dc converter. Moreover, as shown in Fig. 2.34, the status of the inductance current is changed when the load current is varied, and according to (2.1), the additional power from the inductance is transferred to the output side. Thus, (2.12) becomes inaccurate, and the expected transferred current of the unidirectional I^2ACL dc-dc converter also becomes inaccurate. Therefore, the compensation operation should be used after several switching periods until the peak value of the inductance current becomes stable. In addition, in the actual converter system, the measurement noise is unavoidable. To calculate the capacitor charging current i_{Co} with higher accuracy, (2.13) should be further expressed as,

$$k'_{io} = \frac{1}{i_o}[k_{io}i_o - \frac{\sum\limits_{j=1}^{m} \frac{C_o}{T_s}(U_{oj} - U'_{oj})}{m}] \tag{2.14}$$

Based on the compensation method, the influence caused by the efficiency difference can be reduced. Moreover, it can also be employed to reduce the influence caused by middle inductive energy release for unidirectional I²ACL isolated dc-dc converter.

2.2.5 The Implementing Procedures of the Unified FDDC Control Scheme for I²ACL-Type Converter

In previous sections, the general FDDC control scheme is proposed for the I²ACL isolated dc-dc converters. Besides, the design principle of PI parameters and the compensation method for ensuring the performance of the unified method is also presented. Based on these contents, the detailed implementing procedures of the unified FDDC control scheme for arbitrary I²ACL isolated dc-dc converter are demonstrated as shown in Fig. 2.40.

If the simplified circuit of an existing isolated dc-dc converter can be simplified as shown in Figs. 2.7 or 2.18, this converter can be classified as the I²ACL isolated dc-dc converter. Firstly, the modulation method of this converter should be determined, and generally, the phase-shift modulation method is the most suitable modulation method for the I²ACL converter. Based on the employed phase-shift modulation method, the relationship between the transferred current and the phase-shift ratio should be determined, which can be usually found in the existing study [8, 15, 34, 39]. Moreover, to realize the proposed unified FDDC control scheme, the phase-shift ratio should be calculated by the transferred current, which can be employed to connect the outer-loop control value and the phase-shift ratio as shown in Fig. 2.39. Moreover, according to (2.9), the upper limitations of the PI parameters can be determined. Further, the proposed unified FDDC control scheme can be obtained for this specific I²ACL isolated dc-dc converter. Sometimes, the I²ACL dc-dc converter may perform at many different efficiencies under different loading conditions, which influences the dynamic performance of the proposed scheme a little, especially for the unidirectional I²ACL dc-dc converter. So, the presented compensation operation as shown in Sect. 2.2.4 can be employed to reduce this influence, which is based on the disturbance of output-capacitor voltage. Finally, the implementation procedures of the proposed FDDC control scheme can be used to a specific I²ACL dc-dc converter.

2.3   Verification

In this section, by using some popular I²ACL dc-dc converters including the full-bridge dc-dc converter, the DAB dc-dc converter, and the three-phase DAB dc-dc converter as examples, the proposed unified FDDC control strategy is verified.

Fig. 2.40 The implementing procedures of the unified FDDC control scheme for I^2ACL-type isolated dc-dc converter

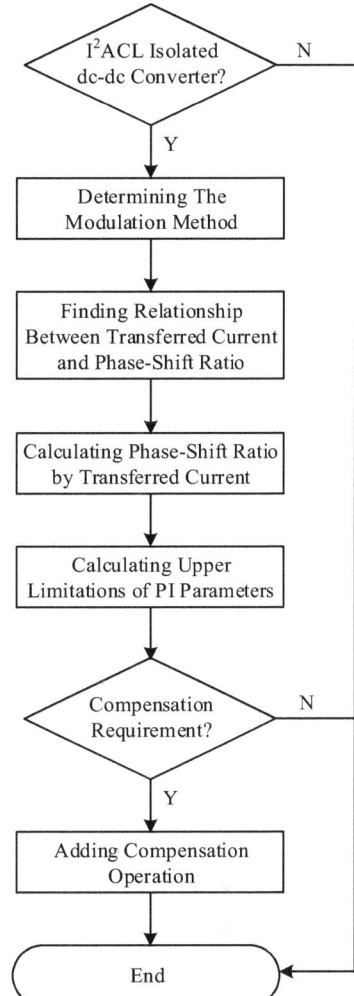

2.3.1 Experiment Results of Full-Bridge DC-DC Converter

For the full-bridge dc-dc converter as shown in Fig. 2.11, the transferred current I_T can be calculated as,

$$I_T = \begin{cases} \frac{(k-1)U_{in}(1-d)^2 T_s}{4nL} & \left(\frac{k-1}{k} < d \leq 1\right) \\ \frac{U_{in}(1-d^2)T_s}{8nL} - \frac{U_o^2 T_s}{8n^3 U_{in}L} & \left(0 \leq d \leq \frac{k-1}{k}\right) \end{cases} \qquad (2.15)$$

According to (2.15), the transferred current I_T of the FB dc-dc converter is not monotone decreasing along with the increasing of the phase-shift ratio d. To design the control

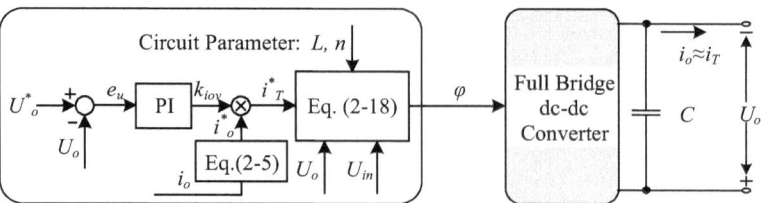

Fig. 2.41 The control block of the FDDC scheme for the full-bridge dc-dc converter

system simply, a middle variable φ can be employed to replace the phase-shift d as,

$$d = 1 - \varphi \tag{2.16}$$

The positive correlation between φ and I_T can be calculated as,

$$I_T = \begin{cases} \frac{(k-1)U_{in}\varphi^2 T_s}{4nL} & \left(0 \le \varphi \le \frac{1}{k}\right) \\ \frac{U_{in}\varphi(2-\varphi)T_s}{8nL} - \frac{U_o^2 T_s}{8n^3 U_{in}L} & \left(\frac{1}{k} < \varphi \le 1\right) \end{cases} \tag{2.17}$$

According to (2.17), the required middle variable φ for the certain transferred current can be shown as,

$$\varphi = \begin{cases} \sqrt{\frac{4nLI_T}{(k-1)U_{in}T_s}} & \left(0 \le I_T \le \frac{(k-1)U_o^2 T_s}{4n^3 LU_{in}}\right) \\ 1 - \sqrt{1 - \frac{8nLI_T}{U_{in}T_s} - \frac{U_o^2}{n^2 U_{in}^2}} & \left(\frac{(k-1)U_o^2 T_s}{4n^3 LU_{in}} < I_T \le \frac{(n^2 U_{in}^2 - U_o^2)U_o T_s}{8n^3 LU_{in}^2}\right) \end{cases} \tag{2.18}$$

Combining Fig. 2.39b, (2.5) and (2.18), the FDDC control scheme for full-bridge dc-dc converter can be shown in Fig. 2.41.

Moreover, the circuit parameters of the full-bridge dc-dc converter can be shown in Table 2.1. As the analysis in Sect. 2.2.1, the middle inductance of the unidirectional I²ACL dc-dc converter releases the storage power during the transient process, which may influence the dynamic performance. So, for the full-bridge dc-dc converter, the experiment results under the FDDC control strategy without or with the compensation operation as shown in Sect. 2.2.4 are provided. Moreover, based on the FDDC control scheme, the corresponding experiment results when the input voltage and the load resistor of the full-bridge dc-dc converter are changed can be shown as Figs. 2.42, 2.43, 2.44, and 2.45, respectively.

When the load resistor R is selected as 40Ω, the experiment results under the FDDC control scheme without compensation operation and the FDDC control scheme with compensation operation when the input voltage U_{in} is changed between 40 and 60 V can be shown in Figs. 2.42 and 2.43 As shown in Fig. 2.42a and 2.43a, when the input voltage is changed, the output-voltage disturbances under the FDDC control scheme without compensation operation are bigger than 0.5 V, and the settling times are obvious. Then, with

Table 2.1 Circuit parameters of the full bridge dc-dc converter

Parameter	Value
Switches	SCT3080
L	50 μH
n	2
f_s	10 kHz
U^*_o	50 V
C_o	1 mF
R	12 Ω or 40 Ω
k_p, k_i	0.05, 0.005

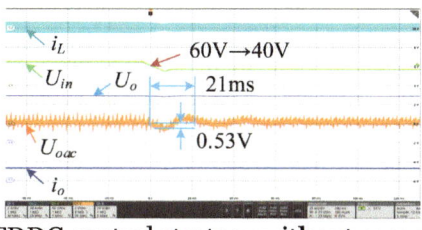

(a). The FDDC control strategy without compensation.

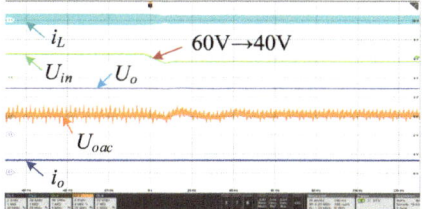

(b). The FDDC control strategy with compensation.

Fig. 2.42 The experiment results when the input voltage is changed from 60 to 40 V. **a, b** (U_{in}: 50 V/div; i_L: 30A/div; U_o: 20 V/div; U_{oac}: 2 V/div; i_o: 2A; t: 20 ms/div)

compensation operation, the output-voltage disturbances under the FDDC control scheme when the input voltage is changed are very small, and the settling times can be omitted as shown in Figs. 2.42b and 2.43b.

In addition, when the input voltage U_{in} is selected as 50 V, the experiment results under the FDDC control scheme without compensation operation and the FDDC control scheme with compensation operation when the load resistor R is changed between 12Ω and 40Ω can be shown in Figs. 2.44 and 2.45. As shown in Figs. 2.44a and 2.45a, when the load resistor is changed, the output-voltage disturbances under the FDDC control scheme without compensation operation are bigger than 0.8 V, and the settling times are obvious. Moreover, with compensation operation, the output-voltage disturbances under

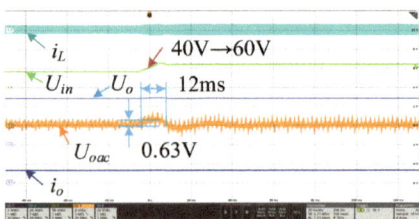

(a). The FDDC control strategy without compensation.

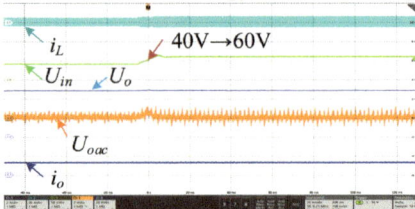

(b). The FDDC control strategy with compensation.

Fig. 2.43 The experiment results when the input voltage is changed from 40 to 60 V. (U_{in}: 50 V/div; i_L: 30A/div; U_o: 20 V/div; U_{oac}: 2 V/div; i_o: 2A; t: 20 ms/div)

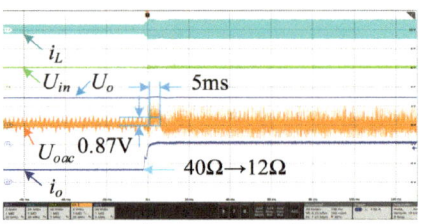

(a). The FDDC control strategy without compensation.

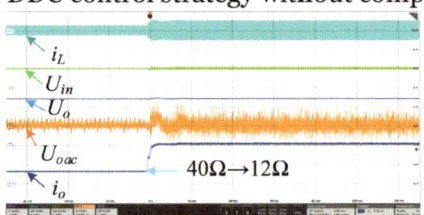

(b). The FDDC control strategy with compensation.

Fig. 2.44 The experiment results when the load resistor is changed from 40 to 12 Ω. (U_{in}: 50 V/div; i_L: 30A/div; U_o: 20 V/div; U_{oac}: 2 V/div; i_o: 2A; t: 20 ms/div)

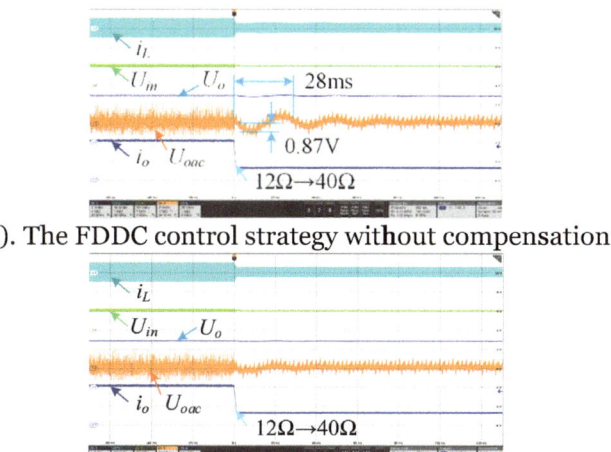

(a). The FDDC control strategy without compensation.

(b). The FDDC control strategy with compensation.

Fig. 2.45 The experiment results when the load resistor is changed from 12 to 40 Ω. (U_{in}: 50 V/div; i_L: 30A/div; U_o: 20 V/div; U_{oac}: 2 V/div; i_o: 2A; t: 20 ms/div)

the FDDC control scheme when the load resistor is changed are very small, and the settling times can be omitted as shown in Figs. 2.44b and 2.45b.

2.3.2   Experiment Results of DAB DC-DC Converter

For the DAB dc-dc converter as shown in Fig. 1.1, the transferred current I_T under the single-phase-shift modulation method can be expressed by the phase-shift ratio D as,

$$I_T = \frac{U_{in}D(1-D)T_s}{2nL} \quad \left(0 \leq D \leq \frac{1}{2}\right) \tag{2.19}$$

The phase-shift ratio D of the DAB dc-dc converter can also be calculated by the transferred current as,

$$\begin{cases} D = \frac{1}{2} - \sqrt{\frac{1}{4} - \frac{2nLI_T}{U_{in}T_s}} & \left(0 \leq I_T \leq \frac{U_{in}T_s}{8nL}\right) \\ D = -\frac{1}{2} + \sqrt{\frac{1}{4} + \frac{2nLI_T}{U_{in}T_s}} & \left(-\frac{U_{in}T_s}{8nL} \leq I_T \leq 0\right) \end{cases} \tag{2.20}$$

Combining Fig. 2.39, (2.5) and (2.20), the FDDC scheme for the DAB dc-dc converter can be shown in Fig. 2.46.

Moreover, the circuit parameters of the DAB dc-dc converter in simulation model can be shown in Table 2.2. Then based on the FDDC control scheme for DAB dc-dc converter, the corresponding simulation results when resistor load, the constant current load (CCL)

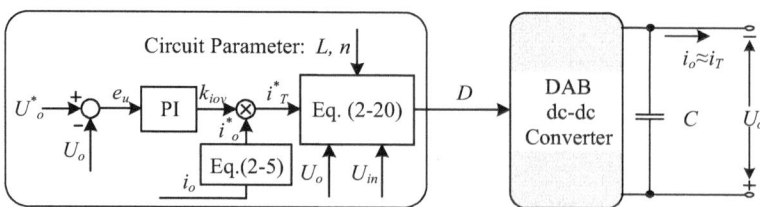

Fig. 2.46 The control block of the FDDC scheme for the DAB dc-dc converter

and constant power load (CPL) of the DAB dc-dc converter are changed can be shown in Fig. 2.47. As shown in Fig. 2.47a, when the resistor load is changed between 10 and 100 Ω, the CCL is changed between 1 and 10 A and the CPL is changed between 0.5 and 5 kW, the output-voltage disturbances are smaller than 1 V as shown in Fig. 2.47b. So, the output voltage can be regarded as maintained at its desired value when the load conditions are changed, and the excellent dynamic performance can be obtained under the proposed FDDC control scheme.

Moreover, the circuit parameters of the DAB dc-dc converter can be shown in Table 2.3. Since DAB dc-dc converter can usually illustrate high efficiency for a large power range, the FDDC scheme is enough for ensuring the fast-dynamic performance when the input voltage and the load resistor are changed. Then based on the FDDC control scheme for DAB dc-dc converter, the corresponding experiment results when the input voltage and the load resistor of the DAB dc-dc converter are changed can be shown in Fig. 2.48.

Table 2.2 Circuit parameters of the DAB dc-dc converter in simulation model

Parameter	Value
Conduction resistor of Switches	30 mΩ
L	80 μH
C	1 mF
n	2
f_s	10 kHz
R	10~100 Ω
U_{in}	200 V
U^*_o	200 V
k_p	$0.05(i_o > 0), -0.05(i_o \leq 0)$
k_i	$0.005(i_o > 0), -0.005(i_o \leq 0)$
CPL	0.5~5 kW
CCL	1~10 A

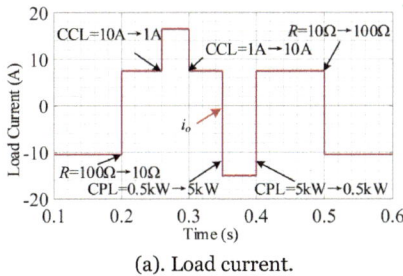
(a). Load current.

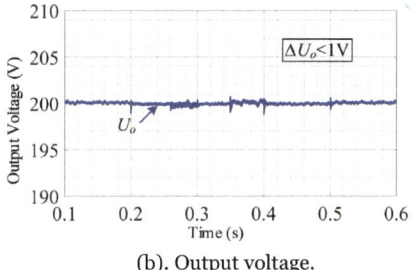
(b). Output voltage.

Fig. 2.47 The simulation result under FDDC control scheme for DAB dc-dc converter

Table 2.3 Circuit parameters of the DAB dc-dc converter in experimental platform

Parameter	Value
Switches	SCT3080KL
L	50 μH
n	2
f_s	40 kHz
R	20~100 Ω
U_{in}	40~50 V
U^*_o	80 V
k_p	0.05
k_i	0.005

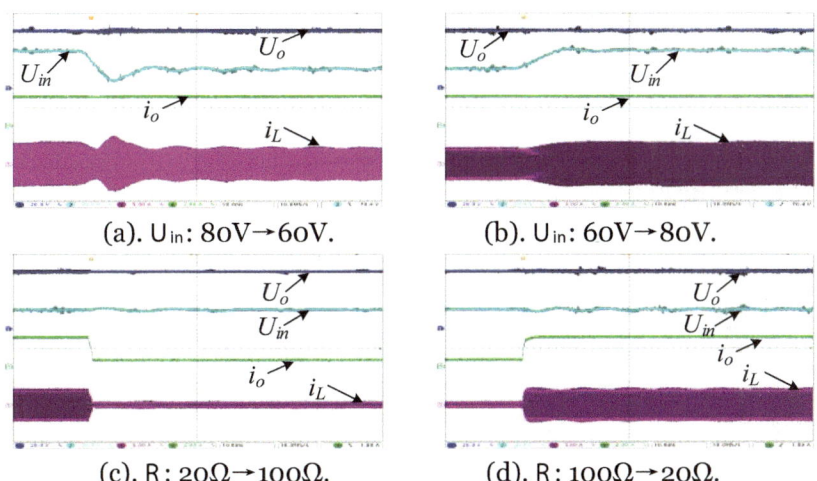

(a). U_{in}: 80V→60V.

(b). U_{in}: 60V→80V.

(c). R: 20Ω→100Ω.

(d). R: 100Ω→20Ω.

Fig. 2.48 The experiment result under FDDC control scheme for DAB dc-dc converter. (U_{in} and U_o: 20 V/div; i_o: 2A/div; i_L: 5A/div; t: 2 ms/div)

When the load resistor R is 20Ω, the experiment results under the FDDC control scheme when the input voltage U_{in} is changed between 60 and 80 V can be shown as Fig. 2.48a and b, where the output voltage can be kept at its desired value. Moreover, when the input voltage U_{in} is 60 V, the experiment results under the FDDC control scheme when the load resistor R is changed between 20 and 100 Ω can be shown as Fig. 2.48c and b, where the output-voltage disturbance can be neglected. Therefore, based on the general FDDC control scheme, an excellent dynamic control scheme can be provided for the DAB dc-dc converter when the input voltage and the load resistor are changed.

2.3.3 Simulation Results of Three-Phase DAB DC-DC Converter

For the three-phase dc-dc converter as shown in Fig. 2.22a, the transferred current I_T under the single-phase-shift modulation method can be expressed by the phase-shift ratio D as,

$$
I_T = \begin{cases} \frac{U_{in}T_s}{2nL}\left(\frac{2}{3} - \frac{D}{2}\right)D & \left(0 \le D < \frac{1}{3}\right) \\ \frac{U_{in}T_s}{2nL}\left[D(1-D) - \frac{1}{18}\right] & \left(\frac{1}{3} \le I_T \le \frac{1}{2}\right) \end{cases} \tag{2.21}
$$

The phase-shift ratio D of the three-phase DAB dc-dc converter can also be calculated by the transferred current as,

$$
D = \begin{cases} \frac{2}{3} - \sqrt{\frac{4}{9} - \frac{4nLI_T}{U_{in}T_s}} & \left(0 \le I_T < \frac{U_{in}T_s}{12nL}\right) \\ \frac{1}{2} - \sqrt{\frac{14}{72} - \frac{2nLI_T}{U_{in}T_s}} & \left(\frac{U_{in}T_s}{12nL} \le I_T \le \frac{7U_{in}T_s}{72nL}\right) \end{cases} \tag{2.22}
$$

Combining Fig. 2.39, (2.5) and (2.22), the FDDC scheme for the three-phase DAB dc-dc converter can be shown in Fig. 2.49.

The circuit parameters of the three-phase DAB dc-dc converter can be shown in Table 2.4. Since three-phase DAB dc-dc converter can usually have high efficiency for a large power range, the FDDC scheme is enough for ensuring the fast-dynamic performance when the input voltage and the load resistor are changed. Moreover, based on the

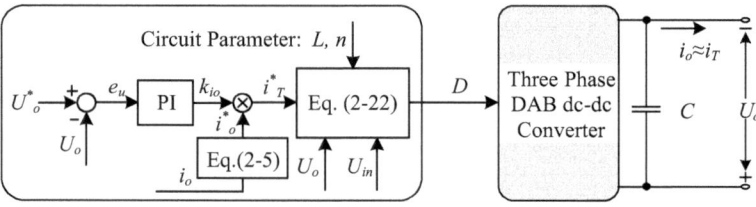

Fig. 2.49 The control block of the FDDC scheme for the three-phase DAB dc-dc converter

Table 2.4 Circuit parameters of the three phase DAB dc-dc converter

Parameter	Value
Switches	SCT3080
L	50 μH
n	1
f_s	10 kHz
U_{in}	100~120 V
U^*_o	100 V
C_o	2 mF
R	12~200 Ω
U_m (measurement noise in voltages)	±0.5 V
R_{on} (conduction resistor in switch)	50 mΩ
k_p, k_i	0.1, 0.01

FDDC control scheme for the three-phase DAB dc-dc converter, the corresponding experiment results when the input voltage and the load resistor of the three-phase DAB dc-dc converter are changed can be shown in Figs. 2.50 and 2.51, respectively.

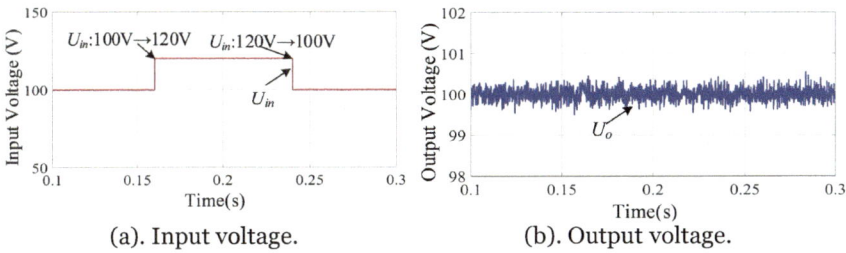

(a). Input voltage. (b). Output voltage.

Fig. 2.50 The simulation result when the input voltage is changed between 100 and 120 V

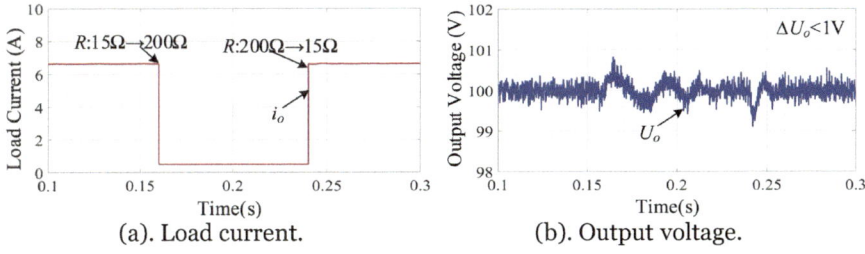

(a). Load current. (b). Output voltage.

Fig. 2.51 The simulation result when the load resistor is changed from 15 to 200 Ω

When the load resistor R is 12 Ω, the simulation results under the FDDC control scheme when the input voltage U_{in} of the three-phase DAB dc-dc converter is changed between 100 to 120 V can be shown as Fig. 2.50, where the output voltage can be kept at its desired value. Moreover, when the input voltage U_{in} is 100 V, the simulation results under the FDDC control scheme when the load resistor R of the three-phase DAB dc-dc converter is changed between 15 and 200 Ω can be shown as Fig. 2.51, where the output-voltage disturbance is smaller than 1 V. Therefore, based on the general FDDC control scheme, an excellent dynamic control scheme can be provided for the three-phase DAB dc-dc converter when the input voltage and the load resistor are changed.

2.4   Summary

In this Chapter, the I²ACL dc-dc converters such as the DAB dc-dc converter, the full bridge dc-dc converter, and their variant topologies are reviewed since these converters have similar transient characteristics. These converters are divided into two groups including the unidirectional type and the bidirectional type. Besides, the current transferred characteristics are analyzed, which reveals the order reducing phenomena of these converters since the intermediary inductance cannot influence the transient performance. Moreover, a unified FDDC control scheme is proposed for providing excellent dynamic performance for this kind of dc-dc converter. In addition, the design principle of PI parameters in the proposed method is presented, and the compensation operation is also provided for ensuring fast-dynamic performance. Notably, since the dynamic equivalence between the DAB dc-dc converter and other I²ACL dc-dc converters have been verified in this Chapter, the control schemes for the DAB-based dc-dc converter systems such as the IPOP, IIOP, IPOS and ISOP DAB dc-dc converter systems can be easily extended to the other I²ACL dc-dc converters with the same modular configurations.

References

1. M. Sharifzadeh, H. Vahedi, A. Sheikholeslami, P. Labbé, and K. Al-Haddad, "Hybrid SHM–SHE Modulation Technique for a Four-Leg NPC Inverter With DC Capacitor Self-Voltage Balancing," *IEEE Transactions on Industrial Electronics,* vol. 62, no. 8, pp. 4890–4899, 2015.
2. D. G. Bandeira, T. B. Lazzarin, and I. Barbi, "High Voltage Power Supply Using a T-Type Parallel Resonant DC–DC Converter," *IEEE Transactions on Industry Applications,* vol. 54, no. 3, pp. 2459–2470, 2018
3. A. Filba-Martinez, S. Busquets-Monge, and J. Bordonau, "Modulation and Capacitor Voltage Balancing Control of Multilevel NPC Dual Active Bridge DC–DC Converters," *IEEE Transactions on Industrial Electronics,* vol. 67, no. 4, pp. 2499–2510, 2020.
4. Y. Li, Y. W. Li, H. Tian, N. R. Zargari, and Z. Cheng, "A Modular Design Approach to Provide Exhaustive Carrier-Based PWM Patterns for Multilevel ANPC Converters," *IEEE Transactions on Industry Applications,* vol. 55, no. 5, pp. 5032–5044, 2019.

5. R. Suryadevara, and L. Parsa, "Full-Bridge ZCS-Converter-Based High-Gain Modular DC-DC Converter for PV Integration With Medium-Voltage DC Grids," *IEEE Transactions on Energy Conversion,* vol. 34, no. 1, pp. 302–312, 2019.

6. X. Kong, and A. M. Khambadkone, "Analysis and Implementation of a High Efficiency, Inter-leaved Current-Fed Full Bridge Converter for Fuel Cell System," *IEEE Transactions on Power Electronics,* vol. 22, no. 2, pp. 543–550, 2007.

7. J. Saeed, and A. Hasan, "Unit Prediction Horizon Binary Search-Based Model Predictive Control of Full-Bridge DC–DC Converter," *IEEE Transactions on Control Systems Technology,* vol. 26, no. 2, pp. 463–474, 2018.

8. N. Hou, and Y. Li, "A Direct Current Control Scheme With Compensation Operation and Circuit-Parameter Estimation for Full-Bridge DC–DC Converter," *IEEE Transactions on Power Electronics,* vol. 36, no. 1, pp. 1130–1142, 2021.

9. N. C. D. Pont, D. G. Bandeira, T. B. Lazzarin, and I. Barbi, "A ZVS APWM Half-Bridge Parallel Resonant DC–DC Converter With Capacitive Output," *IEEE Transactions on Industrial Electronics,* vol. 66, no. 7, pp. 5231–5241, 2019.

10. H. Wu, Y. Lu, T. Mu, and Y. Xing, "A Family of Soft-Switching DC–DC Converters Based on a Phase-Shift-Controlled Active Boost Rectifier," *IEEE Transactions on Power Electronics,* vol. 30, no. 2, pp. 657–667, 2015.

11. Y. Xie, R. Ghaemi, J. Sun, and J. S. Freudenberg, "Model Predictive Control for a Full Bridge DC/DC Converter," *IEEE Transactions on Control Systems Technology,* vol. 20, no. 1, pp. 164–172, 2012.

12. R. Huang, and S. K. Mazumder, "A Soft Switching Scheme for Multiphase DC/Pulsating-DC Converter for Three-Phase High-Frequency-Link Pulsewidth Modulation (PWM) Inverter," *IEEE Transactions on Power Electronics,* vol. 25, no. 7, pp. 1761–1774, 2010.

13. K. Nguyen-Duy, Z. Ouyang, L. P. Petersen, A. Knott, O. C. Thomsen, and M. A. E. Andersen, "Design of a 300-W Isolated Power Supply for Ultrafast Tracking Converters," *IEEE Transactions on Power Electronics,* vol. 30, no. 6, pp. 3319–3333, 2015.

14. Y. Li, F. Li, F. Zhao, and X. You, "Variable-Frequency Control Strategy of Isolated Buck–Boost Converter," *IEEE Journal of Emerging and Selected Topics in Power Electronics,* vol. 7, no. 3, pp. 1824–1836, 2019.

15. D. Sha, D. Chen, S. Khan, and Z. Guo, "Voltage-Fed Three-Phase Semi-Dual Active Bridge DC–DC Converter Utilizing Varying Operating Modes With High Conversion Efficiency," *IEEE Transactions on Power Electronics,* vol. 34, no. 10, pp. 9447–9458, 2019.

16. Z. Li, S. Dusmez, and H. Wang, "A Novel Soft-Switching Secondary-Side Modulated Multiout-put DC–DC Converter With Extended ZVS Range," *IEEE Transactions on Power Electronics,* vol. 34, no. 1, pp. 106–116, 2019.

17. Y. Lu, H. Wu, B. Tu, M. Li, Y. Xia, and Y. Xing, "Ultra-Wide Output Voltage Range Power Supply Based on Modular Switched-Converter Principle," *IEEE Transactions on Power Electronics,* vol. 35, no. 1, pp. 94–106, 2020.

18. A. J. B. Bottion, and I. Barbi, "Input-Series and Output-Series Connected Modular Output Capacitor Full-Bridge PWM DC–DC Converter," *IEEE Transactions on Industrial Electronics,* vol. 62, no. 10, pp. 6213–6221, 2015.

19. S. S. Williamson, A. K. Rathore, and F. Musavi, "Industrial Electronics for Electric Transportation: Current State-of-the-Art and Future Challenges," *IEEE Transactions on Industrial Electronics,* vol. 62, no. 5, pp. 3021–3032, 2015.

20. G. Ortiz, D. Bortis, J. Biela, and J. W. Kolar, "Optimal Design of a 3.5-kV/11-kW DC–DC Converter for Charging Capacitor Banks of Power Modulators," *IEEE Transactions on Plasma Science,* vol. 38, no. 10, pp. 2565–2573, 2010.

21. M. S. Irfan, A. Ahmed, J. Park, and C. Seo, "Current-Sensorless Power-Decoupling Phase-Shift Dual-Half-Bridge Converter for DC–AC Power Conversion Systems Without Electrolytic Capacitor," *IEEE Transactions on Power Electronics,* vol. 32, no. 5, pp. 3610–3622, 2017.
22. W. Jin, F. Z. Peng, J. Anderson, A. Joseph, and R. Buffenbarger, "Low cost fuel cell converter system for residential power generation," *IEEE Transactions on Power Electronics,* vol. 19, no. 5, pp. 1315–1322, 2004.
23. R. W. D. Doncker, D. M. Divan, and M. H. Kheraluwala, "A three-phase soft-switched high power density DC/DC converter for high power applications," *Conference Record of the 1988 IEEE Industry Applications Society Annual Meeting,* 1988, pp. 796–805 vol.1.
24. J. Hu, S. Cui, S. Wang, and R. W. D. Doncker, "Instantaneous Flux and Current Control for a Three-Phase Dual-Active Bridge DC–DC Converter," *IEEE Transactions on Power Electronics,* vol. 35, no. 2, pp. 2184–2195, 2020.
25. H. M. d. O. Filho, D. d. S. Oliveira, and P. P. Praça, "Steady-State Analysis of a ZVS Bidirectional Isolated Three-Phase DC–DC Converter Using Dual Phase-Shift Control With Variable Duty Cycle," *IEEE Transactions on Power Electronics,* vol. 31, no. 3, pp. 1863–1872, 2016.
26. O. Demirel, U. Arifoglu, and K. Kalayci, "Novel three-level T-type isolated bidirectional DC–DC converter," *IET Power Electronics,* vol. 12, no. 1, pp. 61–71, 2019.
27. S. B. Karanki, and D. Xu, "NPC based dual active bridge topology for integrating battery energy storage to utility gird," *2014 IEEE 27th Canadian Conference on Electrical and Computer Engineering (CCECE),* 2014, pp. 1–6.
28. Y. Xuan, X. Yang, W. Chen, T. Liu, and X. Hao, "A Novel NPC Dual-Active-Bridge Converter With Blocking Capacitor for Energy Storage System," *IEEE Transactions on Power Electronics,* vol. 34, no. 11, pp. 10635–10649, 2019.
29. R. M. Burkart, and J. W. Kolar, "Comparative η-ρ-σ Pareto Optimization of Si and SiC Multilevel Dual-Active-Bridge Topologies With Wide Input Voltage Range," *IEEE Transactions on Power Electronics,* vol. 32, no. 7, pp. 5258–5270, 2017.
30. A. Filbà-Martínez, S. Busquets-Monge, and J. Bordonau, "Modulation and capacitor voltage balancing control of a three-level NPC dual-active-bridge DC-DC converter," *IECON 2013 - 39th Annual Conference of the IEEE Industrial Electronics Society,* 2013, pp. 6251–6256.
31. V. N. S. R. Jakka, A. Shukla, and G. Demetriades, "Three-winding transformer based asymmetrical dual active bridge converter," *IET Power Electronics,* vol. 9, no. 12, pp. 2377–2386, 2016.
32. H. Teng, Y. Zhong, and H. Bai, "SiC+ Si three-phase 48 V electric vehicle battery charger employing current-SVPWM controlled SWISS AC/DC and variable-DC-bus DC/DC converters," *IET Electrical Systems in Transportation,* vol. 8, no. 4, pp. 231–239, 2018.
33. N. M. L. Tan, S. Inoue, A. Kobayashi, and H. Akagi, "Voltage Balancing of a 320-V, 12-F Electric Double-Layer Capacitor Bank Combined With a 10-kW Bidirectional Isolated DC–DC Converter," *IEEE Transactions on Power Electronics,* vol. 23, no. 6, pp. 2755–2765, 2008.
34. C. Zhao, S. D. Round, and J. W. Kolar, "An Isolated Three-Port Bidirectional DC-DC Converter With Decoupled Power Flow Management," *IEEE Transactions on Power Electronics,* vol. 23, no. 5, pp. 2443–2453, 2008.
35. V. N. S. R. Jakka, A. Shukla, and G. D. Demetriades, "Dual-Transformer-Based Asymmetrical Triple-Port Active Bridge (DT-ATAB) Isolated DC–DC Converter," *IEEE Transactions on Industrial Electronics,* vol. 64, no. 6, pp. 4549–4560, 2017.
36. W. Song, N. Hou, and M. Wu, "Virtual Direct Power Control Scheme of Dual Active Bridge DC–DC Converters for Fast Dynamic Response," *IEEE Transactions on Power Electronics,* vol. 33, no. 2, pp. 1750–1759, 2018.

37. N. Hou, and Y. Li, "Communication-Free Power Management Strategy for the Multiple DAB-Based Energy Storage System in Islanded DC Microgrid," *IEEE Transactions on Power Electronics,* vol. 36, no. 4, pp. 4828–4838, 2021.

38. V. Blahnik, Z. Peroutka, J. Molnar, and J. Michalik, "Control of primary voltage source active rectifiers for traction converter with medium-frequency transformer," *2008 13th International Power Electronics and Motion Control Conference*, 2008, pp. 1535–1541.

39. J. Hu, Z. Yang, S. Cui, and R. W. D. Doncker, "Closed-Form Asymmetrical Duty-Cycle Control to Extend the Soft-Switching Range of Three-Phase Dual-Active-Bridge Converters," *IEEE Transactions on Power Electronics,* vol. 36, no. 8, pp. 9609–9622, 2021.

The IPOP and IIOP DAB DC-DC Converter Systems

In this chapter, a tunable power sharing control scheme for the IPOP DAB dc-dc converter is proposed in Sect. 3.1. Based on this scheme, excellent dynamic performance can be achieved when the input voltage, the load resistor and the power sharing performance are changed. Besides, the hot-swap operation can also be realized based on the proposed scheme. Moreover, to ensure the desired power sharing performance, the comprehensive CPE schemes proposed for different conditions including the start-up process, the working process, and the plugging-in operation of a new DAB dc-dc converter, respectively. In addition, a communication-free power management strategy is proposed for the IIOP DAB dc-dc converter in Sect. 3.2. This scheme can boost the robustness of the dc-link voltage when the load conditions, the output voltage and the power sharing performance of the power sources are changed. In addition, the hot swap control methods for the IIOP DAB dc-dc converter with only a little influence on output voltage are discussed in detail. Finally, the small-scale simulation model and experimental platform are employed to verify the effectiveness of these proposed schemes for the IPOP DAB dc-dc converter system and the IIOP DAB dc-dc converter system in Sect. 3.3. Then, the logic structure of this Chapter can be summarized in Fig. 3.1.

© The Author(s), under exclusive license to Springer Nature Switzerland AG 2025　　　63
N. Hou, *High-Robust Control Schemes for Dual-Active-Bridge-Based DC–DC Converter Systems in Renewable Energy Applications*, Synthesis Lectures on Power Electronics, https://doi.org/10.1007/978-3-031-72963-8_3

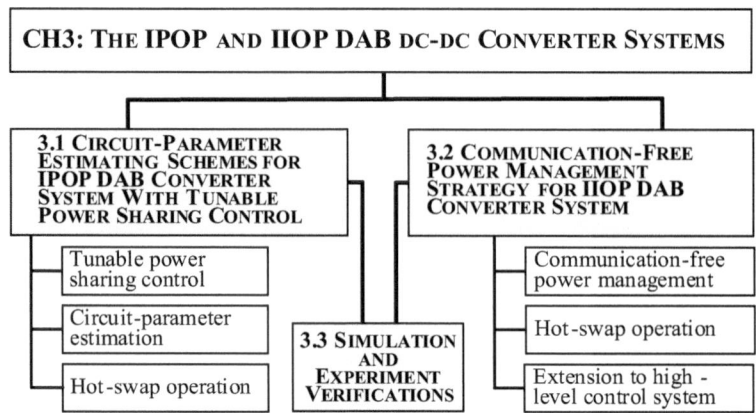

Fig. 3.1 Logic structure of this chapter

3.1   The Comprehensive Circuit-Parameter Estimating Strategies for IPOP DAB DC-DC Converters with Tunable Power Sharing Control

In this section, based on the single-phase-shift (SPS) modulation method, a tunable power sharing control scheme is proposed for the IPOP DAB dc-dc converter system as shown in Fig. 1.7, which is employed to provide the fast-dynamic performance when the input voltage, the load resistor and the power sharing performance among DAB modules are changed. Moreover, to ensure the desired power sharing performance, comprehensive CPE methods are proposed for different conditions including the start-up process, the working process, and the plug-in operation of a new DAB module. In addition, the hot-swap operations of the DAB module are presented with only a little influence on the output voltage.

3.1.1   Analysis of the Tunable Power Sharing Strategy with SPS Modulation Method

To realize power transmission of the DAB dc-dc converter, the SPS modulation method is widely adopted, and this modulation method for each DAB converter can be shown in Fig. 3.2. $S_{1\alpha}$–$S_{8\alpha}$ are square-wave gate driving signals with 50% duty ratio for the corresponding switches of the αth DAB dc-dc module. $U_{ab\alpha}$ and $U_{cd\alpha}$ are output voltages of primary-side and second-side H Bridges, respectively. T_s is the switching period, $i_{L\alpha}$ is the current of the equivalent inductance between two H Bridges, and $i_{o\alpha}$ is the output current of the αth DAB dc-dc module. Moreover, D_α is defined as phase-shift ratio to

Fig. 3.2 The main waveforms of the SPS modulation method

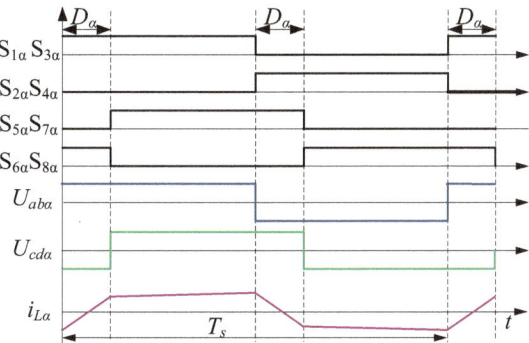

implement power transfer of the corresponding DAB dc-dc converters, and the transferred power P_α can be expressed as,

$$P_\alpha = \frac{n_\alpha U_{in\alpha} U_o D_\alpha (1 - D_\alpha) T_s}{2L_\alpha} \quad (\alpha \in [1, 2, \ldots m]) \tag{3.1}$$

where m is the number of DAB modules, n_α is the transformer turn ratio, and $U_{in\alpha}$ is the input voltage for the αth DAB dc-dc module. When DAB dc-dc converters are connected to the same power source, $U_{in\alpha}$ is the same. Moreover, since this work is focused on the centralized control of IPOP DAB dc-dc system, the switching period is set as the same value for each DAB module. According to (3.1), it is clear that the transferred power of DAB converter can be directly determined by input voltage, output voltage, phase-shift ratio, and circuit parameters including inductance and transformer turn ratio. Assuming reconstructed circuit parameter λ_α for the αth DAB dc-dc converter is equivalent to $2L_\alpha/nT_s$, the transferred power for each DAB converter can be further calculated as,

$$P_\alpha = \frac{U_{in\alpha} U_o D_\alpha (1 - D_\alpha)}{\lambda_\alpha} \tag{3.2}$$

According to (3.2), the required phase-shift ratio D_α for a given transferred power can be determined as,

$$D_\alpha = \frac{1}{2} - \sqrt{\frac{1}{4} - \frac{\lambda_\alpha P_\alpha}{U_{in\alpha} U_o}} \tag{3.3}$$

According to (3.3), the required phase-shift ratio D_α for the given transferred power P_α should be determined by output voltage U_o, input voltage $U_{in\alpha}$ and reconstructed circuit parameter λ_α for each DAB module. Thus, the phase-shift ratio D_α for each DAB converter can be calculated once the required transferred power can be obtained. Ignoring power losses of IPOP DAB dc-dc converter system, the required transferred power can be also expressed as,

$$P_\alpha = k_\alpha P_o^* \tag{3.4}$$

where P_o^* is the required output power of IPOP DAB dc-dc converter system. Therefore, the outer control loop of the flexible power sharing control should offer desired transferred power for obtaining transferred power for each DAB module. Generally, the required output power can be demonstrated as,

$$P_o^* = \frac{U_o^{*2}}{R} \tag{3.5}$$

Combining the measured output voltage and load current, (3.5) can be further expressed as,

$$P_o^* = \frac{i_o U_o^{*2}}{U_o} \tag{3.6}$$

However, the power losses of IPOP DAB dc-dc converter cannot be ignored completely for accurate control of output voltage. Thus, a PI controller should be employed to compensate for the error caused by power losses, and the output value of PI control is named as virtual desired output voltage U_v^*. Then, the required output power in the control system can be described as,

$$P_o^* = \frac{i_o U_o^* U_v^*}{U_o} \tag{3.7}$$

Compared with (3.6) and (3.7), the virtual desired output voltage U_v^* is very close to the desired output voltage U_o^*, since the power losses are always small. Moreover, based on the power control concept, the block diagram of the tunable power sharing strategy can be shown in Fig. 3.3.

As shown in Fig. 3.3, the tunable power sharing strategy can be implemented. In each switching period, the controller system measures the input voltages for each DAB module, the output voltage, and the load current for the IPOP DAB dc-dc converter system. Besides, through the PI controller, the new virtual desired output voltage U_v^* be obtained,

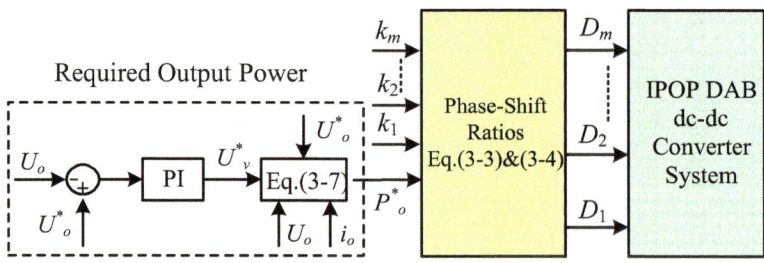

Fig. 3.3 The block diagram of the tunable power sharing strategy

and further combined with (3.7), the required output power P^*_o can be calculated. Moreover, combining required power sharing coefficients k_α and required output power P^*_o, the needed transferred power P_α for each DAB dc-dc converter can be calculated by (3.4). Finally, based on (3.3), the phase-shift ratio D_α for each DAB module can be obtained, and the desired power sharing performance can be achieved.

3.1.2 Analysis of Proposed Circuit-Parameter Estimation Strategies

As illustrated above, the circuit parameters should be used to determine the accurate transferred power for each DAB module. Therefore, inaccurate circuit parameters always result in unexpected power sharing performance for the IPOP DAB dc-dc converter. Therefore, three CPE schemes of the IPOP DAB dc-dc converter are proposed for estimating circuit parameters under different conditions.

A. *The CPE Scheme during Start-Up Process.*

During the start-up process, the circuit parameters should be estimated to correct the previous circuit parameters for each DAB module. When the desired output voltage is achieved, the original power sharing requirement should also be satisfied. Thus, a CPE scheme is proposed for meeting these requirements. Ignoring power losses of IPOP DAB dc-dc system, the total transferred power can also be expressed as,

$$P = \frac{U_o^2}{R} \tag{3.8}$$

The transferred power for each DAB dc-dc converter can be expressed as,

$$P_\alpha = \frac{k_\alpha U_o^2}{R} = \frac{U_{o\alpha} U_o}{R} \tag{3.9}$$

where k_α is the power sharing coefficient, and $U_{o\alpha}$ can be named as the virtual voltage component of the αth DAB dc-dc converter. Therefore, k_α and $U_{o\alpha}$ should meet the relationships as,

$$\begin{cases} \sum_{\alpha=1}^{m} k_\alpha = 1 \\ \sum_{\alpha=1}^{m} U_{o\alpha} = U_o \end{cases} \tag{3.10}$$

Combining (3.3) and (3.9), the phase-shift ratio D_α can be rewritten as,

$$D_\alpha = \frac{1}{2} - \sqrt{\frac{1}{4} - \frac{\lambda_\alpha i_o U_{o\alpha}}{U_{in\alpha} U_o}} \tag{3.11}$$

To determine transferred power for estimating circuit parameters of DAB converter, these DAB modules should be activated one by one sequentially during the start-up process [1]. Besides, combining the circuit parameters of activated DAB dc-dc converters, circuit parameters of the new DAB module can be estimated one by one. Each step has an independent desired output voltage U^*_α, and based on a simple PI controller, the desired output voltage U^*_α can be obtained independently. The output value of PI controller λ_{PI} can be used to calculate circuit parameter for each DAB dc-dc converter. Moreover, to obtain the required power sharing performance of IPOP DAB dc-dc converter when the final desired output voltage U^*_o is achieved, the desired output voltage U^*_α for each step can be calculated as,

$$U^*_\alpha = (k_1 + k_2 \ldots + k_i)U^*_o \tag{3.12}$$

During the start-up process of IPOP DAB dc-dc system, the first DAB converter is activated, and when U^*_1 is achieved, λ_1 can be obtained as λ_{PI}. Then, the second DAB module is activated, and the desired output voltage is U^*_2, when the error u_e between desired output voltage and output voltage is zero, λ_2 can be obtained as λ_{PI}-λ_1. Similarly, the estimated circuit parameter λ_α for each DAB dc-dc converter can be shown as,

$$\begin{cases} \lambda_\alpha|_{U_o=U^*_\alpha} = \lambda_{PI}|_{U_o=U^*_\alpha} - \lambda_{\alpha-1}|_{U_o=U^*_{\alpha-1}} \ldots - \lambda_1|_{U_o=U^*_1} \\ \lambda_1 = \lambda_{PI}|_{U_o=U^*_1} \end{cases} \tag{3.13}$$

Combining (3.9), (3.11), (3.12) and (3.13), the block diagram of the CPE strategy of IPOP DAB dc-dc converter can be shown in Fig. 3.4.

As shown in Fig. 3.4, during the start-up process, DAB dc-dc converters are activated in sequence, and when the total desired output voltage is obtained, the circuit parameters of each DAB module and the required original power sharing performance of IPOP DAB dc-dc converter can be obtained at the same time. To distinguish the moments when each desired output voltages are reached, the least square method [2] is adopted, and the sum

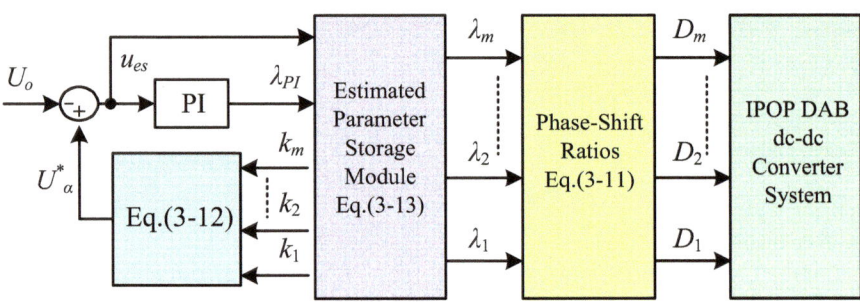

Fig. 3.4 The block diagram of CPE scheme during start-up process

of squared voltage error u_{es} can be expressed as,

$$u_{es} = \sum_{j=1}^{g} u_{ej}^2 \tag{3.14}$$

where g is the adopted number of voltage error, and u_{es} is the sum of squared voltage error. When u_{es} is smaller than the stored limit value, the desired output voltage for each step can be treated as acquisition in controller system. When the desired output voltage for IPOP DAB dc-dc converter is reached, the original power sharing requirement can be obtained according to the desired output voltage settings for each step, and the circuit parameters for IPOP DAB dc-dc converter can be also estimated.

In addition, when the control method is switched from the CPE strategy to the tunable power sharing scheme, the original value of PI controller in the tunable power sharing scheme should be determined by the eventual conditions of circuit-parameter estimation strategy. Thus, the original value of PI controller U^*_v can be depicted as,

$$U^*_v = \frac{\sum_{i=1}^{m} U_{o\alpha} U_o}{U^*_o} \tag{3.15}$$

In (3.15), the original value of U^*_v is obtained by using the final condition under circuit-parameter estimation strategy of IPOP DAB dc-dc converter. Based on (3.15), when the control method is switched from circuit-parameter estimation strategy to tunable power sharing control, phase-shift ratio for each DAB module is not changed suddenly and significantly. Therefore, the switching operation between these two methods can be very smooth.

B. *The CPE Scheme during Working Condition.*

Sometimes, the inductance of DAB dc-dc converter may be changed during working condition, especially when the inductance temperature is changed [3]. Therefore, the corresponding circuit-parameter estimating scheme should also be discussed [4, 5]. To address this issue, a CPE strategy with a seamless switching concept is proposed, and the core principle is based on the power compensation between the reference module and the other module. Assuming the circuit parameter of the first DAB modulation is employed as the base, and the reconstructed circuit parameter λ_1 can be expressed as,

$$\lambda_1 = \frac{2L_1}{n_1 T_s} \tag{3.16}$$

To determine the relationship between the circuit parameter of the first DAB dc-dc module and the circuit parameter of the other DAB dc-dc module, the pre-disturbance of phase-shift ratio for the first DAB module can be expressed as ΔD_1. Thus, the variation of transferred power can be expressed as,

$$\Delta P_1 = \frac{U_o}{\lambda_1}[U_{in1}(D_1 - \Delta D_1)(1 - D_1 + \Delta D_1) - U_{in1}D_1(1 - D_1)] \tag{3.17}$$

The expected power sharing coefficient variation Δk_1 of the first DAB module can be calculated as,

$$\Delta k_1 = \frac{\Delta P_1}{P_o^*} = \frac{U_o U_{in1}(2D_1 \Delta D_1 - \Delta D_1 - \Delta D_1^2)}{\lambda_1 P_o^*} \tag{3.18}$$

Generally, the input voltage of the first DAB module and the load resistor may be changed during the estimating process, the phase-shift ratio of the first DAB module should be modified for meeting these variations. Assuming the phase-shift ratio is D'_1, the input voltage is U'_{in1} and the load resistor is R' when the desired output voltage is reached again, the relationship between transferred power of the first DAB module with the new input voltage and the new load resistor can be illustrated as,

$$\frac{U_o U_{in1}(D_1 - \Delta D_1)(1 - D_1 + \Delta D_1)}{\lambda_1} = \frac{R'}{R}\frac{U_o U'_{in1}D'_1(1 - D'_1)}{\lambda_1} \tag{3.19}$$

According (3.19), (3.17) can be modified as,

$$\Delta P_1 = \frac{U_o}{\lambda_1}\left[\frac{R'}{R}U'_{in1}D'_1(1 - D'_1) - U_{in1}D_1(1 - D_1)\right] \tag{3.20}$$

A certain DAB converter can be used to compensate this decrease of power sharing coefficient Δk_1. When the desired output voltage is achieved again, the expected compensation transferred power mapped to the original condition of this adopted DAB dc-dc module can be calculated as,

$$\Delta P_\beta = -\Delta P_1 = \frac{U_o}{\lambda'_\beta}\left[\frac{R'}{R}U'_{in\beta}D'_\beta(1 - D'_\beta) - U_{in\beta}D_\beta(1 - D_\beta)\right] \tag{3.21}$$

Here D'_β is the new phase-shift ratio, λ'_β is the estimated circuit parameter and $U'_{in\beta}$ is the new input voltage of the adopted DAB converter. Combining (3.17) and (3.21), λ'_β can be expressed as,

$$\lambda'_\beta = -\frac{[R'U'_{in\beta}D'_\beta(1 - D'_\beta) - RU_{in\beta}D_\beta(1 - D_\beta)]\lambda_1}{[R'U'_{in1}D'_1(1 - D'_1) - RU_{in1}D_1(1 - D_1)]} \tag{3.22}$$

According to (3.22), the circuit parameter of the adopted DAB converter can be estimated again. During the estimating process, the phase-shift ratio of the first DAB module can be calculated as,

$$D_1 = \frac{1}{2} - \sqrt{\frac{1}{4} - \frac{\lambda_1 P_o^*(k_1 + \Delta k_1)}{U_{in1} U_o}} \tag{3.23}$$

The phase-shift ratio of the adopted DAB module for compensating this transferred power change of the first DAB module can be calculated as,

$$D_\beta = \frac{1}{2} - \sqrt{\frac{1}{4} - \frac{\lambda_\beta P_o^* k_\beta'}{U_{in2} U_o}} \tag{3.24}$$

where k_β' is the virtual power sharing coefficient for this DAB module. During the estimated condition, k_β' can be adjusted by the PI controller. In addition, the phase-shift ratios for other DAB dc-dc converters can be calculated as,

$$D_\alpha = \frac{1}{2} - \sqrt{\frac{1}{4} - \frac{\lambda_\alpha P_\alpha}{U_{in\alpha} U_o}} \quad (\alpha \in [2, 3, 4 \ldots m] \ \& \ \alpha \neq \beta) \tag{3.25}$$

Similarly, the circuit parameter of all DAB modules can be also acquired by using the circuit-parameter reference of the first DAB module as,

$$\lambda'_{\alpha=2,3\ldots} = -\frac{[R'U'_{in\alpha}D'_\alpha(1 - D'_\alpha) - RU_{in\alpha}D_\alpha(1 - D_\alpha)]\lambda_1}{[R'U'_{in1}D'_1(1 - D'_1) - RU_{in1}D_1(1 - D_1)]} \tag{3.26}$$

According to (3.26), the CPE method for IPOP DAB dc-dc converter under working condition can be implemented. R and R' can be calculated as the quotient of the output voltage U_o and the load current i_o at the corresponding time. Moreover, the power sharing coefficient for these DAB modules can be obtained as,

$$\begin{cases} k_1'' = k_1 \\ k'_{\alpha=2,3\ldots} = \frac{\lambda_{\alpha=2,3\ldots}}{\lambda'_{\alpha=2,3\ldots}} k'_{\alpha=2,3\ldots} \end{cases} \tag{3.27}$$

Combining (3.23), (3.24), (3.25) and (3.26), the block diagram of CPE scheme during working condition can be illustrated as Fig. 3.5.

As shown in Fig. 3.5, the CPE scheme during working condition can be implemented. Assuming the circuit parameter of the first DAB module as reference, the circuit parameter of other DAB modules can be estimated one by one by finding the relationship between the circuit parameter of the first DAB module and the circuit parameter of the estimated DAB converter. To reduce disturbance of output voltage, ΔD_1 should be adjusted slowly. When all circuit parameters are determined, the core control should be switched to the tunable power sharing scheme, and the estimated circuit parameters should also be used. To switch these two strategies smoothly, the new desired transferred power $P_o^{*'}$ and the actual power sharing coefficient k_α''' should be calculated as,

Fig. 3.5 The block diagram of CPE scheme during working condition

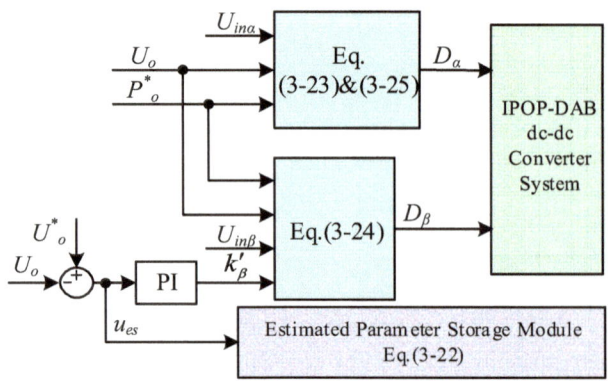

$$\begin{cases} k_\alpha''' = \dfrac{k_\alpha''}{\sum_{\alpha=1}^{m} k_\alpha''} \\ P_o^* = P_o^* \left(k_1 + \displaystyle\sum_{\alpha=2}^{m} k_\alpha'' \right) \end{cases} \tag{3.28}$$

According to (3.28), the sum of actual power sharing coefficients can equivalent to 1 again. Thus, the phase-shift ratio for each DAB converter can be expressed as,

$$\begin{cases} D_1 = \frac{1}{2} - \sqrt{\frac{1}{4} - \frac{\lambda_1 k_1''' P_o^{*\prime}}{U_{in1} U_o}} \\ D_\alpha = \frac{1}{2} - \sqrt{\frac{1}{4} - \frac{\lambda_\alpha' k_\alpha''' P_o^{*\prime}}{U_{in\alpha} U_o}} \quad (\alpha \in [2,3,4\ldots m]) \end{cases} \tag{3.29}$$

According to (3.29), the required power sharing performance can be obtained. Then, the tunable power sharing strategy can be employed to allocate the power sharing proportion for the IPOP DAB dc-dc converter. In addition, the circuit parameter λ_1 of the first DAB dc-dc converter is employed to estimate the circuit parameters of other DAB dc-dc converters. Therefore, when this circuit parameter is inaccurate, the estimating performance of this CPE scheme may be affected. In terms of the variation of λ_1, there are two conditions including the variation before estimating process and the variation during estimating process. Since only a short time is required for this proposed CPE method, the circuit parameter of this first DAB dc-dc converter can be regarded as constant value. Thus, the proposed CPE scheme will not be affected.

Moreover, when the circuit parameter λ_1' of the first DAB dc-dc converter is changed before the estimating process, the CPE method for working condition cannot also be affected. Assuming the phase-shift variation of the first DAB dc-dc converter is the same, the new estimated circuit parameters λ_β'' of other DAB dc-dc converters can be expressed as,

$$\lambda_\beta'' = -\frac{[R'R'U_{in\beta}'D_\beta'(1-D_\beta') - RU_{in\beta}D_\beta(1-D_\beta)]\lambda_1'}{[R'U_{in1}'D_1'(1-D_1') - RU_{in1}D_1(1-D_1)]} \tag{3.30}$$

Assuming $\lambda'_1 = k_\lambda \lambda_1$, , (3.30) can be further illustrated as,

$$\lambda''_\beta = -\frac{[R'U'_{in\beta}D'_\beta(1-D'_\beta) - RU_{in\beta}D_\beta(1-D_\beta)]k_\lambda \lambda_1}{[R'U'_{in1}D'_1(1-D'_1) - RU_{in1}D_1(1-D_1)]} \tag{3.31}$$

In addition, according to (3.3), when the circuit parameters λ'_α are inaccurate, the actually transferred power of the DAB dc-dc converters can be shown as,

$$P_\alpha = \frac{\lambda'_\alpha k_\alpha P^*_o}{\lambda_\alpha} \quad (\alpha \in [1,2,3\ldots m]) \tag{3.32}$$

Combining (3.22), (3.31) and (3.32), the power sharing performance of this output-parallel DAB dc-dc converter can be expressed as,

$$\begin{aligned} P_2 : P_3 \ldots P_\alpha \ldots : P_m &= \frac{\lambda'_1 k_1}{\lambda_1} : \frac{\lambda''_2 k_2}{\lambda_2} \ldots \frac{\lambda''_\alpha k_\alpha}{\lambda_\alpha} \ldots : \frac{\lambda''_m k_m}{\lambda_m} \\ &= \frac{k_\lambda \lambda_1 k_1}{\lambda_1} : \frac{k_\lambda \lambda'_2 k_2}{\lambda_2} \ldots \frac{k_\lambda \lambda'_\alpha k_\alpha}{\lambda_\alpha} \ldots : \frac{k_\lambda \lambda'_m k_m}{\lambda_m} \\ &\approx k_1 : k_2 \ldots k_\alpha \ldots : k_m \end{aligned} \tag{3.33}$$

Therefore, based on (3.33), when the circuit parameter of the first DAB dc-dc converter is not accurate, the estimating performance under the proposed CPE scheme is the same as when the circuit parameter is accurate, and the estimating performance will not be affected for the IPOP DAB dc-dc converter system.

C. The CPE Scheme for a New Plugged-in DAB Module.

Sometimes, when a new DAB dc-dc converter should be plugged in, the circuit parameter of this new module may be inaccurate. Therefore, a corresponding CPE scheme is required to determine this newly reconstructed circuit parameter λ_{new}, which can be expressed as,

$$\lambda_{new} = \frac{2L_{new}}{n_{new}T_s} \tag{3.34}$$

here L_{new} is the inductance, n_{new} is the transformer turn ratio of the new DAB dc-dc converter. The block diagram of CPE scheme for a new plugged-in DAB dc-dc converter can be shown in Fig. 3.6.

As shown in Fig. 3.6, the desired output power for the m DAB dc-dc converters should be changed as the product of the total required output power of the IPOP DAB dc-dc converter and the sum of power sharing coefficients of the m DAB modules. Notably, when the new DAB dc-dc converter is plugged in, the virtual desired output voltage U^*_v should remain unchanged. Then, the tunable power sharing control strategy can be employed until the steady-state condition of IPOP DAB dc-dc converter is achieved again by using the least square method. Moreover, the estimated circuit parameter λ_{new} of this

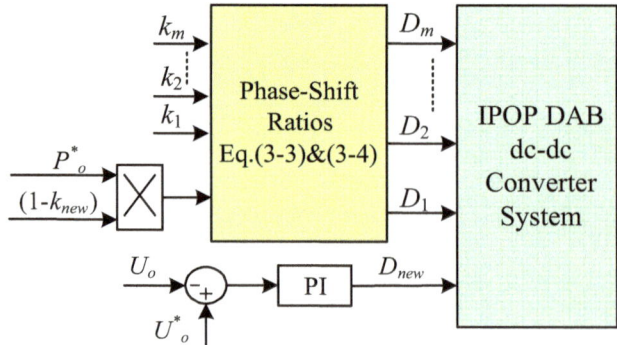

Fig. 3.6 The block diagram of CPE method for a new plugged-in DAB module

DAB module can be expressed as,

$$\lambda_{new} = \frac{U_{innew} U_o D_{new}(1 - D_{new})}{k_{new} P_o^*} \tag{3.35}$$

here U_{innew} is input voltage, D_{new} is phase-shift ratio, k_{new} is power sharing coefficient of this new DAB dc-dc converter in steady state condition. Different from the circuit-parameter estimation strategy, the virtual desired output voltage U_v^* should not be changed when the tunable power sharing control scheme is employed after the plug-in process, since the virtual desired voltage U_v^* is always close to the desired output voltage U_o^*.

D. *The Extension of These Proposed Schemes for Other Phase-Shift Methods.*

When the power-level control scheme is employed for the DAB dc-dc converter, it is super easy to extend this scheme to some advanced phase-shift methods such as the dual-phase-shift method [6], the extended-phase-shift method [7], and the triple-phase-shift method [8]. A hybrid static and dynamic optimizing scheme [9] which is based on the power-based control [9] and the triple-phase-shift method [10] has verified this characteristic.

Based on the minimum-current-stress modulation method for DAB dc-dc converter [9, 11], the corresponding phase-shift ratios $D_{\alpha 1} \sim D_{\alpha 3}$ can be calculated by the corresponding transferred power P_α for each DAB dc-dc converter as shown in Table 3.1, where p_α is the unified transferred power. According to (3.3), (3.11) and (3.23)~(3.25), these proposed schemes including tunable power sharing and CPE strategies are always calculating the transferred power firstly. Therefore, once the transferred power P_α is obtained, these proposed schemes can be implemented with the minimum-current-stress modulation method according to Table 3.1. In addition, in the CPE scheme for a new plugged-in DAB dc-dc converter as shown in Fig. 3.6, the output value of the PI controller can be replaced by the unified transferred power directly since the change of output value of the PI controller is positively correlated with the transferred power of the DAB dc-dc converter.

3.1.3   Analysis of Hot Plug-Out or Plug-In Processes

When one DAB dc-dc converter should be plugged, the power sharing coefficient of this module can be regarded as zero. The main requirements are keeping the continuity of inductance current and reducing the influence of the energy stored in inductance on the output voltage. Generally, for plugging out of DAB dc-dc module, the procedure can be divided into two steps. In the first step, the transferred power of this DAB converter should be controlled to zero, and then, when the energy of inductance is consumed completely by the conducting resistor, switches of this DAB dc-dc converter can be turned off to reduce power consumption caused by the gate driving circuits. The schematic diagram of this plug-out process can be shown in Fig. 3.7.

As shown in Fig. 3.7, during the first step, $S_1 S_4 S_5 S_8$ should be turned on and $S_2 S_3 S_6 S_7$ should be turned off. Thus, the output voltages of two H Bridges in DAB converter are zero. Based on (3.1), the transferred power of this DAB dc-dc converter is zero. Moreover, there is a close loop for the flowing current of inductance through $S_1 S_4 S_5 S_8$. The equivalent circuit of this step can be depicted in Fig. 3.8.

In Fig. 3.8, R_{on} is the equivalent resistor of the conducting loop for inductance current. When the energy in inductance is consumed by this conducting resistor, switches of DAB dc-dc converter can be turned off. Assuming the change of inductance current is linear, the time duration of step 1 can be calculated as,

$$t_h >= \frac{\frac{1}{2} L i_L^2}{R_{on} \left(\frac{1}{2} i_L\right)^2} = \frac{2L}{R_{on}} \tag{3.36}$$

Therefore, in order to guarantee the complete consumption of inductance energy, the time duration should meet the requirement as expressed in (3.36). Then, as shown in Fig. 3.7, switches of this plug-out DAB module can be turned off in the second step.

Moreover, when a DAB module with accurate circuit parameters should be plugged in, the tunable power sharing scheme can be employed to configure the new power sharing coefficients for each DAB dc-dc converter. In addition, if the circuit parameter of the new DAB dc-dc converter is unknown, the CPE scheme presented for a new plugged-in DAB module should be used to determine the circuit parameter for the new DAB dc-dc converter.

Table 3.1 Optimized solutions of minimum-current-stress modulation method with transferred power

Voltage conditions	Unified transferred power	Range of p	Middle variable	Phase-shift ratio
$k > 1$	$p_\alpha = \frac{8L_\alpha P_\alpha}{n_\alpha U_{in\alpha} U_o T_s}$	$0 \leq p_\alpha < 2\frac{k_\alpha - 1}{k_\alpha^2}$	$D_{\alpha 1} = 1 - \sqrt{\frac{p_\alpha}{2(k_\alpha - 1)}}$	$\begin{cases} D_{\alpha 2} = (k_\alpha - 1)(1 - D_{\alpha 1}) \\ D_{\alpha 3} = D_{\alpha 1} \end{cases}$
		$2\frac{k_\alpha - 1}{k_\alpha^2} \leq p_\alpha \leq 1$	$D_{\alpha 1} = (k_\alpha - 1)\sqrt{\frac{1 - p_\alpha}{k_\alpha^2 - 2k_\alpha + 2}}$	$\begin{cases} D_{\alpha 2} = \frac{k_\alpha - 2}{2(k_\alpha - 1)}D_{\alpha 1} + \frac{1}{2} \\ D_{\alpha 3} = \frac{k_\alpha - 2}{2(k_\alpha - 1)}D_{\alpha 1} + \frac{1}{2} \end{cases}$
$k \leq 1$		$0 \leq p_\alpha < 2(k_\alpha - k_\alpha^2)$	$D_{\alpha 1} = 1 - \sqrt{\frac{p_\alpha}{2k_\alpha(1 - k_\alpha)}}$	$\begin{cases} D_{\alpha 2} = 0 \\ D_{\alpha 3} = k_\alpha D_{\alpha 1} - k_\alpha + 1 \end{cases}$
		$2(k_\alpha - k_\alpha^2) \leq p_\alpha \leq 1$	$D_{\alpha 2} = \frac{1}{2}\left(1 - \sqrt{\frac{1 - p_\alpha}{2k_\alpha^2 - 2k_\alpha + 1}}\right)$	$\begin{cases} D_{\alpha 1} = 0 \\ D_{\alpha 3} = 2k_\alpha D_{\alpha 2} - D_{\alpha 2} - k_\alpha + 1 \end{cases}$

Fig. 3.7 The schematic diagram of the plug-out procedure

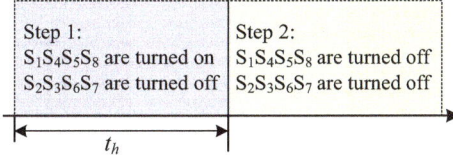

Step 1:	Step 2:
$S_1S_4S_5S_8$ are turned on	$S_1S_4S_5S_8$ are turned off
$S_2S_3S_6S_7$ are turned off	$S_2S_3S_6S_7$ are turned off

t_h

Fig. 3.8 The equivalent circuit for the first step in plug-out process

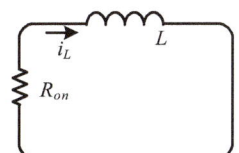

i_L L

R_{on}

3.2 Communication-Free Power Management Strategy for the IIOP DAB DC-DC Converter System

In this section, based on the droop control concept, a communication-free power management strategy is proposed for the IIOP DAB dc-dc converter system as shown in Fig. 1.9, which is used to realize the fast-dynamic performance when the input voltage, the load resistor and the power sharing performance among DAB modules are changed. Moreover, the hot-swap operations of the DAB module are presented for plugging-in or plugging-out the DAB module. In addition, the potential extension to low-bandwidth high-level control system of the proposed strategy is presented.

3.2.1 The Communication-Free Power Management Strategy

For the IIOP DAB dc-dc converter System, the communication-free power management strategy is proposed to maintain the dc-link voltage when the input voltage, the load condition, and the power sharing performance of the ESS are varied. Moreover, the design principle of the PI parameters is presented for ensuring the stability of the proposed scheme.

A. *The Proposed Communication-Free Power Management Strategy.*

To realize flexible power transmission, the SPS modulation method is the most popular modulation method for the DAB dc-dc converter. Thus, the SPS modulation method is adopted, which can be illustrated in Fig. 3.9, where $S_{\alpha 1}{\sim}S_{\alpha 8}$ are the switching signals for the corresponding switches, $U_{ab\alpha}$ is the output voltage of the primary-side H Bridge, $U_{cd\alpha}$ is the output voltage of the secondary-side H Bridge, $i_{L\alpha}$ is the inductance current,

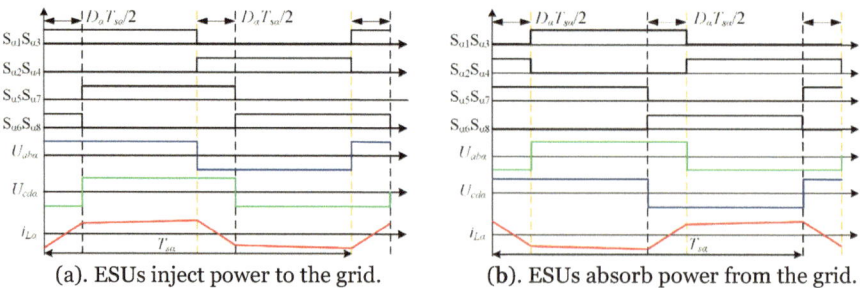

(a). ESUs inject power to the grid.   (b). ESUs absorb power from the grid.

Fig. 3.9 The SPS modulation method of DAB converter for bidirectional power flowing conditions

D_α is the phase-shift ratio and $T_{s\alpha}$ is the switching period of the αth DAB dc-dc converter for the αth ESU.

When the ESU injects power into the dc grid, the transferred power of the DAB module is assumed as positive, and when the ESU absorbs power from the dc grid, the transferred power of the DAB module is assumed as negative. According to Fig. 3.9, the transferred power under the SPS modulation method can be expressed as,

$$P_\alpha = \begin{cases} \frac{U_{in\alpha}U_{dc}D_\alpha(1-D_\alpha)T_{s\alpha}}{2n_\alpha L_\alpha} & (P_\alpha \geq 0) \\ -\frac{U_{in\alpha}U_{dc}D_\alpha(1-D_\alpha)T_{s\alpha}}{2n_\alpha L_\alpha} & (P_\alpha < 0) \end{cases} \tag{3.37}$$

According to (3.37), the phase-shift ratio D_α can be calculated as,

$$D_\alpha = \begin{cases} \frac{1}{2} - \sqrt{1 - \frac{8L_\alpha P_\alpha}{n_\alpha U_{in\alpha}U_{dc}T_{\alpha s}}} & (P_\alpha \geq 0) \\ \frac{1}{2} - \sqrt{1 + \frac{8n_\alpha L_\alpha P_\alpha}{U_{in\alpha}U_{dc}T_{\alpha s}}} & (P_\alpha < 0) \end{cases} \tag{3.38}$$

Moreover, to implement the communication-free control performance, the droop control concept is adopted, and the desired dc-link voltage $U_{dc\alpha}$ for each EUS can be expressed as,

$$U_{dc\alpha} = U_{nom} - \frac{P_\alpha}{k_\alpha P_T^*} = U_{nom} - \frac{P_\alpha}{k_\alpha U_{nom}i_o^*} \tag{3.39}$$

where k_α is the droop coefficient, U_{nom} is the nominal voltage of the dc grid, P_T^* is the total desired output power and i_o^* is the desired output current of the ESS. In (3.39), the total desired output power P_T^* is employed to unify the transferred power P_α of the αth ESU, and the desired output current i_o^* of the ESS can be expressed as,

$$i_o^* = \frac{U_{nom}}{R_{TE}} = \frac{U_{nom}i_o}{U_{dc}} \tag{3.40}$$

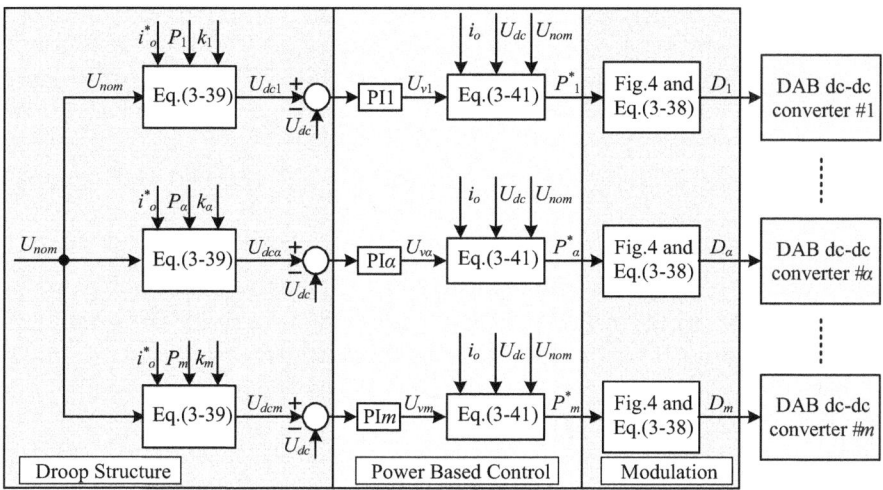

Fig. 3.10 The communication-free power management strategy for IIOP DAB dc-dc converter system with ESUs

where R_{TE} is the total equivalent load resistor. In addition, based on the power control concept, the required transferred power for the αth DAB module can be shown as,

$$P_\alpha^* = \frac{U_{v\alpha} U_{nom} i_o}{U_{dc}} \qquad (3.41)$$

where $U_{v\alpha}$ is named as the virtual dc-link voltage which is the output value of the outer-loop PI controller. Combining (3.37)~(3.41) and Fig. 3.9, the communication-free power management scheme can be illustrated in Fig. 3.10.

In Fig. 3.10, the proposed communication-free power management strategy can be realized for the IIOP DAB dc-dc converter System. For each ESU, the output current i_o, the input voltage $U_{in\alpha}$ of the energy storage component and the dc-link voltage U_{dc} are measured at the beginning of each switching period. Besides, based on (3.40), the desired output current i_o^* of the ESS is calculated, and based on (3.37), the transferred power of each ESU can be obtained. Since the load current is adopted, the excellent dynamic response can be obtained by using the power-based control in this proposed communication-free power management strategy. Further, combining (3.39), the desired dc-link voltage $U_{dc\alpha}$ for each ESU can be obtained. Moreover, based on the power control concept, the required transferred power P_α^* for each ESU can be acquired by (3.41). In addition, combing Fig. 3.9 and (3.38), the phase-shift ratio D_α can be calculated for realizing the required transferred power for each ESU. Since the input voltage is acted as the feedback value for each modulation structure, the required transferred power from the power-based control can be ensured even when the input voltage of DAB module is changed. Therefore, the fast-dynamic performance can be achieved when the output voltage of energy storage component is changed. Importantly, other phase-shift modulation methods such as the dual-phase-shift modulation method [6], the extended-phase-shift modulation method [7, 12] and the triple-phase-shift modulation method [10, 13] can be employed for boosting the efficiency of the whole converter system since the transferred power is acted as the middle control value between the power-based control structure and the modulation structure [14].

Therefore, based on the proposed communication-free power management strategy for ESS, the ultrafast dynamic response can be obtained to ensure the stability of the dc system. In addition, since the control loop for each ESU contains the PI controller based on the droop control concept, which can boost the autonomy characteristic of the ESU, and it is easy to realize the hot swap of the ESU. In addition, according to (3.39), when the droop parameter of one ESU is reduced for smaller transferred power, other ESUs will share more power with the same droop parameters, so the steady-state dc-link voltage will be close to the nominal dc-link voltage. Conversely, the steady-state dc-link voltage will be away to the nominal dc-link voltage.

B. *The Design Principle of the PI Parameter and the Droop Coefficients.*

For DAB dc-dc converter, the relationship between the phase-shift ratio and the transferred power or current is similar in steady-state condition or during the transient process [14]. Thus, leakage inductances of these converters do not affect the dynamic performance, and DAB modules can be directly regarded as current sources. Then, IIOP DAB dc-dc converter System can be simplified as Fig. 3.11.

Generally, when input voltage or load condition are changed, the desired voltages $U_{dc\alpha}$

from the droop control structure can be treated as constant values. Thus, the power-based control method for the IIOP DAB dc-dc converter system can be illustrated in Fig. 3.12. Assuming ESUs are working on power balancing performance, the transfer function of each DAB module can be expressed as,

$$H_\alpha(s) = \frac{U_{nom}i_o}{mU_{dc}^2}\frac{k_{p\alpha}S + k_{i\alpha}}{S}\frac{1}{SC_{dc}} \approx \frac{i_o}{mU_{dc}}\frac{k_{p\alpha}S + k_{i\alpha}}{S}\frac{1}{SC_{dc}} \tag{3.42}$$

Fig. 3.11 The simplified circuit of IIOP DAB dc-dc converter System

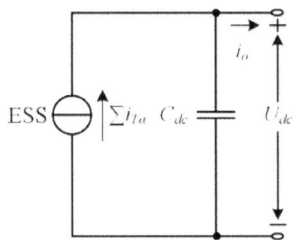

Fig. 3.12 The power-based control scheme for the DAB dc-dc converter

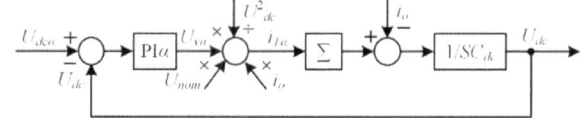

where m is the number of the ESU, $k_{p\alpha}$ is the proportionality coefficient and $k_{i\alpha}$ is the integral coefficient of the PI controller. According to (3.42), since DAB dc-dc converter is the first-order system with capacitive character, the phase margin is always close to $90°$. Then, to ensure the stability of DAB module, the cross-over frequency is set as the switching frequency $f_{s\alpha}$. Moreover, assuming $k_{p\alpha}$ is ten times as $k_{i\alpha}$, $k_{p\alpha}$ can be calculated as,

$$k_{p\alpha} = \frac{2\pi f_{s\alpha} C_{dc} m U_{dc}}{i_o} = 2\pi f_{s\alpha} C_{dc} m R_{TE} \qquad (3.43)$$

Assuming the switching frequency is 10 kHz, the bode diagram can be demonstrated as Fig. 3.13.

Based on (3.43), $k_{p\alpha}$ is usually bigger than 100, and combining Fig. 3.13, the control system can provide a stable dc-link voltage. However, although oscillations of the dc-link voltage U_{dc} can be easily evitable with the dc-link capacitor, there are obvious disturbances in phase-shift ratio with measurement noise since the power transferring range of DAB converter is limited [14]. Thus, the disturbances ΔD_α of the phase-shift ratio caused by the measurement noises should also be treated as a criterion to evaluate the stability of the DAB dc-dc converter. Besides, combining (3.38), (3.41) and Fig. 3.10, and assuming the measurement noise U_{dcmn} and the difference of the measurement noises between two successive switching periods are close, the phase-shift ratio disturbances ΔD_α can be expressed as,

Fig. 3.13 The bode diagram of the power-based control method for the DAB module

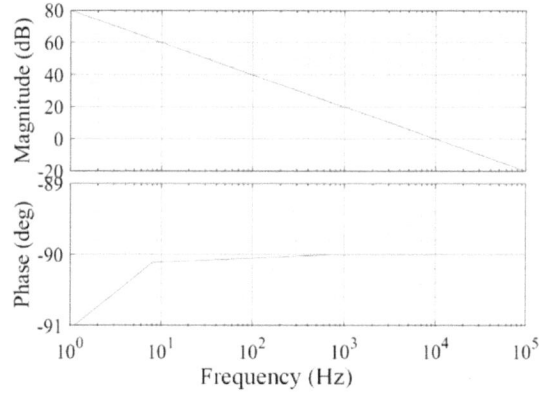

$$\Delta D_\alpha \approx \sqrt{1 - \frac{8 n_\alpha L_\alpha [U_{\alpha v} + (k_{p\alpha} + k_{i\alpha}) U_{dcmn}] U_{nom} i_o}{U_{in\alpha} U_{dc}^2 T_{\alpha s}}}$$
$$- \sqrt{1 - \frac{8 n_\alpha L_\alpha U_{\alpha v} U_{nom} i_o}{U_{in\alpha} U_{dc}^2 T_{\alpha s}}} \qquad (3.44)$$

According to (3.44) and ignoring the higher minimum term, PI parameters can be expressed as,

$$(k_{p\alpha} + k_{i\alpha}) \approx \left| \frac{\Delta D_\alpha}{U_{dcmn}} \sqrt{1 - \frac{8 n_\alpha L_\alpha U_{\alpha v} U_{nom} i_o}{U_{in\alpha} U_{dc}^2 T_{\alpha s}}} \frac{U_{in\alpha} U_{dc}^2 T_{\alpha s}}{4 U_{nom} i_o n_\alpha L_\alpha} \right| \qquad (3.45)$$

Further, assuming the maximum phase-shift ratio disturbance is $\Delta D_{\alpha max}$, (3.45) can also be expressed as,

$$(k_{p\alpha} + k_{i\alpha}) \leq \left| \frac{\Delta D_{\alpha max}}{U_{dcmn}} \sqrt{1 - \frac{8 n_\alpha L_\alpha U_{\alpha v} U_{nom} i_o}{U_{in\alpha} U_{dc}^2 T_{\alpha s}}} \frac{U_{in\alpha} U_{dc}^2 T_{\alpha s}}{4 U_{nom} i_o n_\alpha L_\alpha} \right| \qquad (3.46)$$

$k_{i\alpha}$ can be designed as the tenth of $k_{p\alpha}$ since transferred power in steady-state condition is mainly dependent on the feedback values, and (3.46) can be further expressed as,

$$\begin{cases} k_{p\alpha} \leq \left| \frac{\Delta D_{\alpha max}}{U_{dcmn}} \sqrt{1 - \frac{8 n_\alpha L_\alpha U_{\alpha v} U_{nom} i_o}{U_{in\alpha} U_{dc}^2 T_{\alpha s}}} \frac{U_{in\alpha} U_{dc}^2 T_{\alpha s}}{4 U_{nom} i_o n_\alpha L_\alpha} \right| \\ k_{i\alpha} \leq \left| \frac{\Delta D_{\alpha max}}{10 U_{dcmn}} \sqrt{1 - \frac{8 n_\alpha L_\alpha U_{\alpha v} U_{nom} i_o}{U_{in\alpha} U_{dc}^2 T_{\alpha s}}} \frac{U_{in\alpha} U_{dc}^2 T_{\alpha s}}{4 U_{nom} i_o n_\alpha L_\alpha} \right| \end{cases} \qquad (3.47)$$

In addition, the droop coefficient k_α is also the main control parameter in the communication-free power management strategy for power sharing performance of different ESUs. Based on the droop control concept, the voltage error ΔU_{dc} between the steady-state dc-link voltage U_{dc} and the nominal dc-link voltage U_{nom} in the communication-free power management system can be shown in Fig. 3.14. Moreover, according to (3.39), the voltage error ΔU_{dc} can be expressed as,

$$\Delta U_{dc} = U_{nom} - U_{dc} = \frac{P_\alpha}{k_\alpha U_{nom} i_o^*} \qquad (3.48)$$

When the power balancing performance is realized among different ESUs for the ESS, (3.48) can be further illustrated as,

$$\Delta U_{dc} = \frac{1}{k_\alpha m} \qquad (3.49)$$

Assuming the allowed maximum voltage error is ΔU_{dcmax}, the droop coefficient k_α can be further expressed as,

Fig. 3.14 The regulation characteristic of the droop control in the communication-free power management strategy

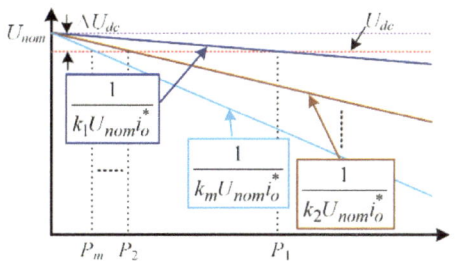

$$k_\alpha \geq \frac{1}{m \Delta U_{dc\,\text{max}}} \tag{3.50}$$

Notably, the allowed maximum voltage error ΔU_{dcmax} between the nominal dc-link voltage and the steady dc-link voltage should be bigger than the measurement noise U_{dcmn}. Thus, the droop coefficient k_α can be further expressed as,

$$\frac{1}{m \Delta U_{dc\,\text{max}}} \leq k_\alpha < \frac{1}{m U_{dcmn}} \tag{3.51}$$

Usually, in order to reduce the impact of the measurement noise obvious, (3.51) can be further expressed for the practice application as,

$$\frac{1}{m \Delta U_{dc\,\text{max}}} \leq k_\alpha \leq \frac{1}{2m U_{dcmn}} \tag{3.52}$$

3.2.2 The Plug-In or Plug-Out Processes of the DAB Module

The plug-in or plug-out operations of the ESU are required for repairing the energy storage equipment and extending the power capacity for the ESS. Based on this proposed communication-free power management strategy, the new ESU can be easily plugged in with only a minor influence on the dc-link voltage since the PI controller can provide a buffer function. Therefore, the transferred power of the new plugged-in ESU can be increased slowly, and based on the adjusting function of the PI controller and droop controller in other ESUs, the steady-state condition of the ESS can be obtained when the new ESU is completely plugged in. According to (3.49), when the number of the ESUs is increased, the steady-state dc-link voltage is a little close to the nominal dc-link voltage, and the voltage error is preferred to be smaller. Thus, when the steady-state condition is obtained again, the actual dc-link voltage will be closer to the nominal dc-link voltage.

In addition, when the βth DAB-based ESU should be plugged out, the transferred power of this ESU should become zero firstly, and the plug-out process of this DAB-based ESU can be shown in Fig. 3.15, where the virtual output voltage $U_{v\beta}$ is gradually reduced to zero, and the transferred power of this ESU can be decreased to zero. Notably, this plugged-out ESU can be treated as load by other ESUs, and the βth DAB-based ESU will not control the dc-link voltage actively. Moreover, with the feedback value of the load current and the input voltage, this plugged-out ESU can offer a timely response when the load condition or input voltage are changed. Thus, the robustness of the dc-link voltage can be ensured during the plug-out process of the ESU. In addition, when the transferred power of the βth DAB-based ESU becomes zero, the storage energy in the leakage inductance of the transformer should be consumed before plug-out action, and with the parallel diodes, these storage energies can transfer to the ESU and the dc-link bus by turning off all the switches. The corresponding circuit can be shown as Fig. 3.16. Further, when the inductance current becomes zero, there is no exchanging power between the ESU and the dc-link bus and the flowing current in DAB dc-dc module, and the βth DAB-based ESU can be completely plugged-out from the ESS. According to (3.49), when the number of the ESUs is decreased, the voltage error between the actual dc-link voltage and the nominal voltage is preferred to be bigger. Thus, when the steady-state condition is obtained again, the actual dc-link voltage will be a little away from the nominal dc-link voltage in the isolated dc microgrid.

Fig. 3.15 The plug-out process of the βth DAB-based ESU

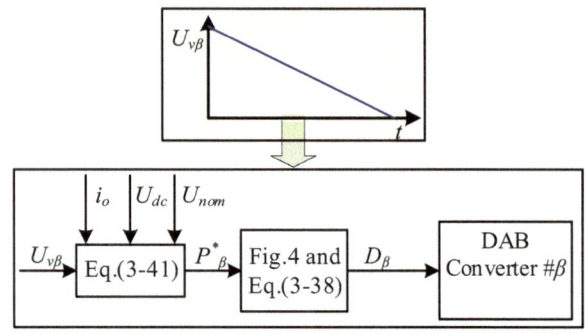

Fig. 3.16 The circuit condition for consuming the storage energies in the inductances of the DAB dc-dc module

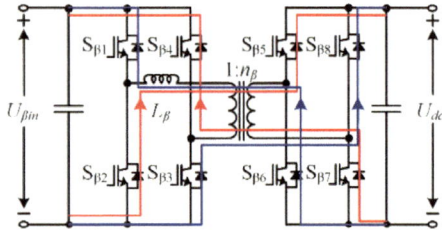

3.2.3 The Potential Extension to Low-Bandwidth High-Level Control System

By presetting different droop coefficients for different state of charge conditions of the energy storage equipment, the balanced state of charge performance among different ESUs can be obtained through a relatively long-time fuzzy regulation function. However, sometimes, higher requirement of the state of charge performance of the ESUs should be provided, and the centralized man–machine interaction system of the ESS may be required. Thus, a high-level control system may be required, and the potential system structure for the communication-free power management strategy with high-level central controller can be shown in Fig. 3.17.

In Fig. 3.17, the central controller can be employed to ensure advanced state of charge of different ESUs and provide good human–machine interaction system of the whole ESS. Importantly, based on the proposed communication-free power management strategy, the high-level central control system will not affect the dc-link voltage, and when the ESU is plugged in or plugged out, the reprogramming operation is not required since each ESU has self-regulating ability with the integrated close-loop structure containing the independent PI controller for adjusting the dc-link voltage.

3.3 Verification

In this section, based on the small-scale experiment platforms, the experiment results are employed to verify the effectiveness of the proposed schemes for the IPOP DAB dc-dc converter system and the IIOP DAB dc-dc converter system. Moreover, for the IIOP DAB dc-dc converter system, the simulation results are also provided for monitoring some middle control values of the proposed communication-free power management strategy.

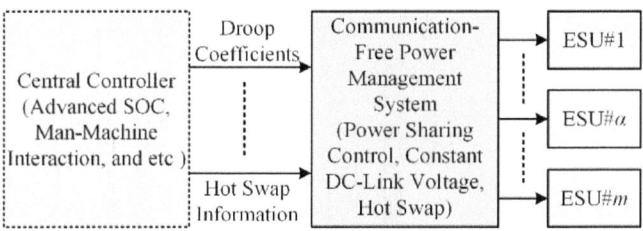

Fig. 3.17 The potential system structure for the proposed communication-free power management strategy with low-bandwidth high-level central controller

3.3.1 The Comprehensive Circuit-Parameter Estimating Strategies with Tunable Power Sharing Control for IPOP DAB DC-DC Converter System

Based on an experimental platform, these presented methods including the circuit-parameter estimation strategy, the tunable power sharing control and the plug-in and plug-out methods are verified for the IPOP DAB dc-dc converter system. Moreover, the dynamic response under the tunable power sharing scheme is also illustrated. The main parameters of the IPOP DAB dc-dc converter system are illustrated in Table 3.2. Notably, each DAB module in the IPOP DAB converter system share the same input voltage in the experimental section.

A. *The Circuit-Parameter Estimation Strategy during Start-up Process.*

When load resistor R is selected as 12 Ω and required power sharing coefficients k_1 and k_2 both equal to 0.5, Fig. 3.18 shows experimental results during the start-up process and the corresponding steady waveforms of IPOP DAB dc-dc converter system. As shown in Fig. 3.18a, the DAB modules are activated one by one, where when output voltage is reached at 30 V, the second DAB converter is activated. Moreover, when the total desired output voltage is achieved, the steady state waveforms can be described in Fig. 3.18b. Besides, according to (3.1), the transferred power of the first DAB module P_1 is 160 W and the transferred power of the second DAB module P_2 is 170 W. The actual power sharing coefficients of the two DAB modules are 0.485 and 0.515, respectively. Compared with the required power sharing coefficients and the actual ones, the desired power sharing performance of IPOP DAB dc-dc converter can be obtained even without circuit-parameter knowledge.

Table 3.2 Experimental parameters of the IPOP DAB dc-dc converter system

Parameter	Value
Number of DAB modules	2
Switches	SCT3080
L_1	400 µH
L_2	447 µH, 200 µH
n_1 and n_2	2
f_s	10 kHz
R	12 or 18 Ω
U_{in1} and U_{in2}	80 to 90 V
U^*_o	60 V

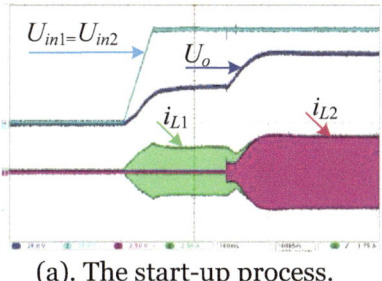
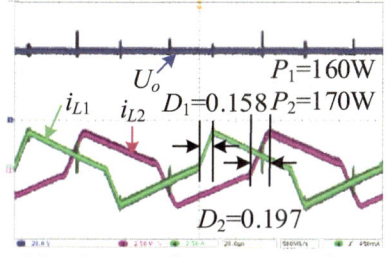

(a). The start-up process.   (b). The steady state waveforms.

Fig. 3.18 Experimental results during start-up process and steady state waveforms. (U_{in} and U_o: 20 V/div; i_{L1} and i_{L2}: 2.5 A/div; Time: 100 ms/div (**a**), 20 μs/div (**b**))

B. *The Circuit-Parameter Estimation Strategy during Working Condition.*

When load resistor R is selected as 18 Ω, input voltage U_{in} is selected as 80 V, inductance L_2 is changed into 200 μH, and k_1 and k_2 are both equivalent to 0.5. Firstly, the same inductance values for these two DAB dc-dc modules are adopted as 400 μH in DSP controller, and then, the CPE strategy during working condition is adopted to estimate L_2. When the estimating process is finished, the core control method is switched into the tunable power sharing scheme. Figure 3.19 shows the estimating process.

Fig. 3.19 Experimental results during adopting CPE scheme during working condition

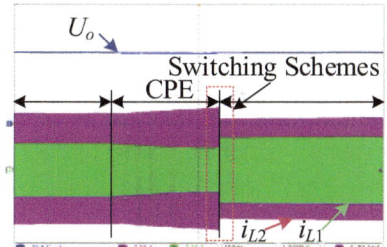

(a). Waveforms during estimating process.

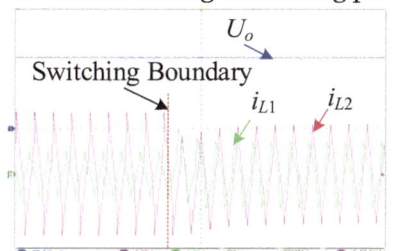

(b). Zoom-in waveforms for switching from CPE scheme to tunable power sharing strategy.

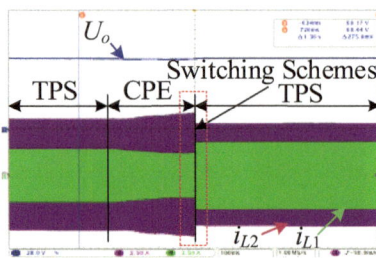

Fig. 3.20 Experimental results during adopting CPE scheme with inaccurate circuit-parameter reference during working condition

As shown in Fig. 3.19a, the output voltage U_o can keep stable during the estimating process and the scheme switching process. Moreover, according to (3.28), when the estimating process is finished, the core control strategy can be switched into tunable power sharing strategy in a switching period (see Fig. 3.19b). Moreover, the same inductance values for these two DAB dc-dc modules are adopted as 447 µH not 400 µH in the DSP controller. Besides, the CPE strategy during working condition is adopted to estimate L_2, and Fig. 3.20 shows the corresponding estimating process. As shown in Fig. 3.20, the output voltage U_o can also be regarded as stable, the CPE scheme can be switched into the tunable power sharing strategy smoothly, and the inductance-current values of i_{L1} and i_{L2} are very similar to these values shown in Fig. 3.19a.

In addition, Fig. 3.21 illustrates the steady-state waveforms before and after estimating process. As shown in Fig. 3.21a, the desired power sharing performance cannot be achieved with wrong mismatch value of the second DAB module, and the transferred power of the second module is twice as much as that of the first module. Then, when the estimating process is finished, the required power sharing coefficients of the two DAB modules can be nearly obtained as 0.506 and 0.494 (see Fig. 3.21b).

C. *The Tunable Power Sharing Control.*

For the same system, when the input voltage is changed from 80 to 90 V after estimating procedure, k_1 is changed to 0.667 and k_2 is changed to 0.333. Then, k_1 and k_2 are both set to 0.5 eventually to illustrate additional transients. Figure 3.22 shows experiment results of transient waveforms for IPOP DAB dc-dc converter with 12 Ω load resistor. As depicted in Fig. 3.22, when the power sharing coefficients are changed, the output voltage can stay stable by using the presented tunable power sharing control, and the corresponding power sharing performances can be obtained from Fig. 3.23. When k_1 is set to 0.667 and k_2 is set to 0.333, the steady state waveforms are shown as Fig. 3.23a. Similarly, the transferred power of the first DAB module P_1 is 212 W and the transferred power of the second DAB module P_2 is 108 W. So, the actual power sharing coefficients of

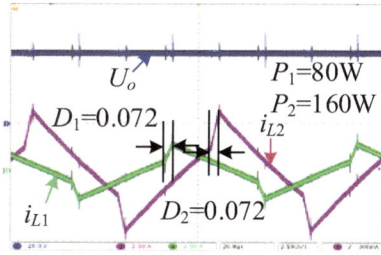

(a). The steady state waveforms before estimating scheme.

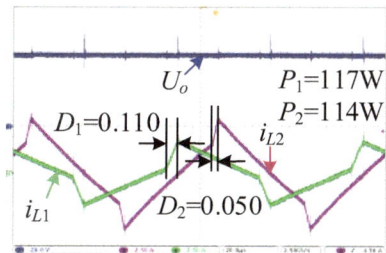

(b). The steady state waveforms after estimating strategy.

Fig. 3.21 Steady state experimental results before and after estimating process

Fig. 3.22 Experimental results when power sharing coefficients are changed. (U_{in} and U_o: 20 V/div; i_o: 2.5 A/div; Time: 1 s/div)

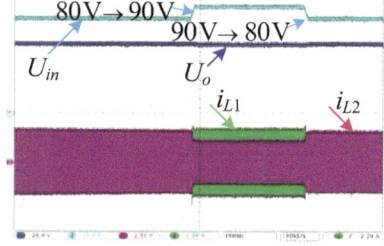

the two DAB modules are 0.663 and 0.338, respectively, and the desired power sharing performance can be achieved. Further, when the input voltage is varied to 80 V again, the desired power sharing requirement is returned to original condition, and the steady state waveforms can be shown in Fig. 3.23b. Compared with Figs. 3.18b and 3.23b, the original power sharing requirement of IPOP DAB dc-dc converter can be also achieved.

D. *The Hot Plug-in and -out Technology.*

When R is selected as 18 Ω, only the first DAB module is worked first, and when R is changed to 12 Ω, the second DAB module is also activated to participate in power transfer with $k_1 = k_2 = 0.5$. Based on the circuit-parameter estimation method during

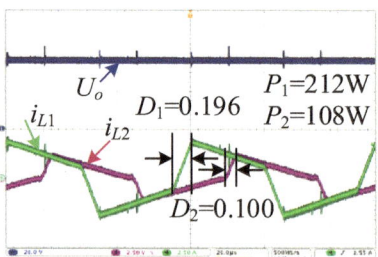

(a). The steady state waveforms when k_1=0.667 and k_2=0.333.

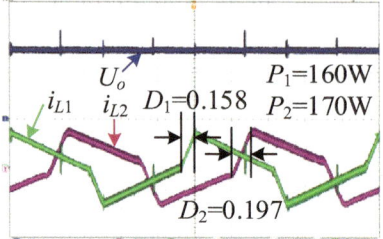

(b). The steady state waveforms when k_1=0.5 and k_2=0.5.

Fig. 3.23 Experimental results of inductance currents with different power sharing coefficients. (U_o: 20 V/div; i_{L1} and i_{L2}: 2.5 A/div; Time: 20 μs/div)

plug-in DAB dc-dc converter, the experiment results of the plug-in process can be shown in Fig. 3.24. As shown in Fig. 3.24a, a few disturbances of output voltage are noticed, but the desired output voltage can be obtained quickly. Besides, according to Fig. 3.24b, the transferred powers of the two DAB dc-dc converters can be obtained as 165 W and 170 W, respectively. So, the actual power sharing coefficients of the two DAB modules are 0.493 and 0.507, respectively, proving the desired power sharing performance.

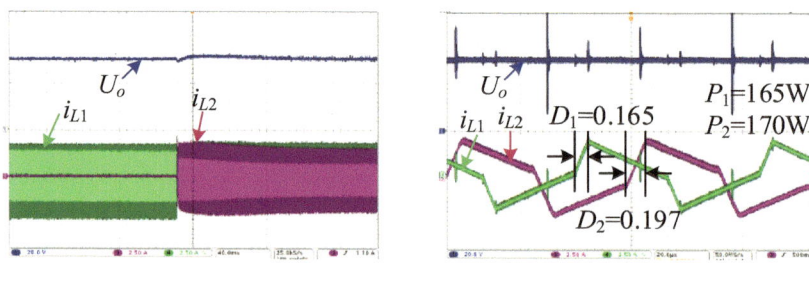

(a). The plug-in process.   (b). The steady state waveforms.

Fig. 3.24 Experimental results under circuit-parameter estimation during plug-in DAB module. (U_{in} and U_o: 20 V/div; i_{L1} and i_{L2}: 2.5 A/div; Time: 40 ms/div (**a**), 20 μs/div (**b**))

When R is changed from 12 to 18 Ω, the first DAB dc-dc converter is plugged out based on the presented plug-out method for IPOP DAB dc-dc converter, and the corresponding experiment result can be shown in Fig. 3.25. As shown in Fig. 3.25a, the inductance current of the first DAB converter can turn into zero, and then, the switching signal of this DAB converter can be set as 0 quickly. Moreover, the output voltage can remain as its desired value. In addition, when R is changed from 18 to 12 Ω again, the first DAB module is plugged in to share power transmission as shown in Fig. 3.25b. It is obvious that the output voltage can stay stable during this plug-in process based on the presented tunable power sharing control by reallocating the power sharing coefficients for each DAB converter. Similarly, the equal power sharing performance ($k_1 = k_2 = 0.5$) of these two DAB modules can be also obtained when the first DAB converter is plugged in again, which can be seen from Fig. 3.26.

E. *The Dynamic Performances under Disturbances of Input Voltage and Load Resistor.*

When k_1 and k_2 are both set to 0.5, Fig. 3.27 shows experiment results with the variated input voltage between 80 and 90 V and variated load resistor between 12 and 18 Ω. As shown in Fig. 3.27, under input voltage is changed or load resistor variations, the excellent dynamic performances can be obtained without any disturbances of output voltage for the

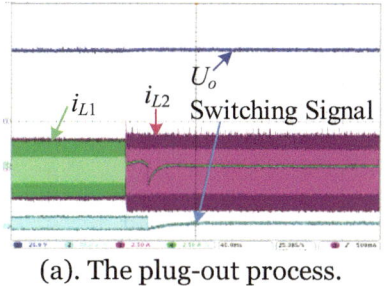

(a). The plug-out process.   (b). The plug-in process.

Fig. 3.25 Experimental results during plug-out DAB module. (U_{in} and U_o: 20 V/div; i_{L1} and i_{L2}: 2.5 A/div; Time: 40 ms/div)

Fig. 3.26 Experimental results of inductance currents when the first DAB module is plugged in again. (U_{in} and U_o: 20 V/div; i_{L1} and i_{L2}: 2.5 A/div; Time: 20 μs/div)

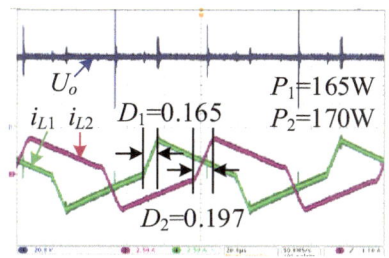

IPOP DAB dc-dc converter system. The steady state results under different conditions are shown in Fig. 3.28. Similarly, the desired power sharing performance for IPOP DAB dc-dc converter under different situations can be obtained under the tunable power sharing control strategy.

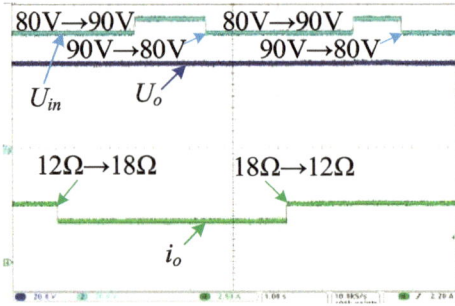

Fig. 3.27 Experimental results under disturbances of input voltage and load resistor. (U_{in} and U_o: 20 V/div; i_o: 2.5 A/div; Time: 1 s/div)

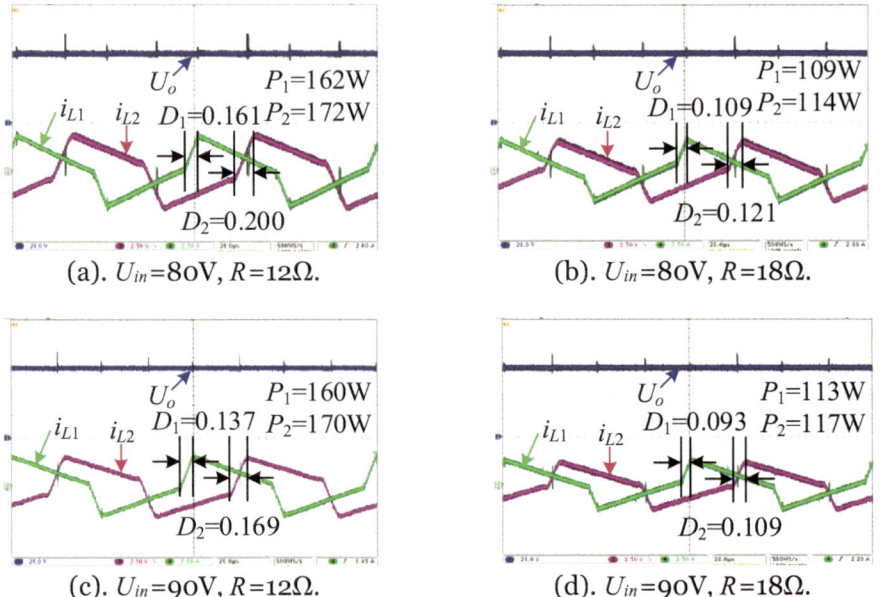

Fig. 3.28 Experimental results of inductance currents under different conditions. (U_o: 20 V/div; i_{L1} and i_{L2}: 2.5 A/div; Time: 20 μs/div)

3.3.2 Communication-Free Power Management Strategy for the IIOP DAB DC-DC Converter System

In this section, based on Simulink model and experiment platform of the IIOP DAB-based dc-dc converter system, the effectiveness of the proposed communication-free power management strategy will be verified when the input voltage, the load condition and the droop coefficient of the ESS are changed, and when the ESU is plugged-in or -out, which can be employed to simulate the potential conditions in the practice application.

A. *Simulation Results for the Communication-Free Power Management Strategy.*

The Simulink model with three DAB-based dc-dc converter system is established, and the circuit parameters are illustrated in Table 3.3.

When the droop coefficients k_1 k_2 and k_3 are 0.5, the simulation results when the input voltage U_{in1} of the first DAB module and the load resistor R_{TE} are changed can be shown in Fig. 3.29. As shown in Fig. 3.29c, when the input voltage U_{in1} of the first DAB converter and the load condition R_{TE} are varied (see Fig. 3.29a and b), the corresponding transferred power can be quickly obtained by using the proposed strategy, and the transferred power of these three ESU are the same. So, as shown in Fig. 3.29d, the disturbances of dc-link voltage can be neglected.

When the original droop coefficients k_1 k_2 and k_3 are 0.5, the load resistor R_{TE} is 10Ω, the input voltage U_{in1} is 60 V, the input voltage U_{in2} is 60 V and the input voltage U_{in3} is 50 V, the simulation results when the droop coefficient k_1 of the first ESU is changed between 0.5 and 1 can be illustrated in Fig. 3.30. As shown in Fig. 3.30a, when k_1 is changed, the desired dc-link voltage of the first ESU is changed suddenly. Based on the adjusting function of the PI controller and the droop controller, the new steady state of the dc-link voltage can be acquired. When k_1 is increased, the new steady dc-link voltage

Table 3.3 Simulation parameters of the IIOP DAB dc-dc converter system

Parameter	Value
Number of DAB Modules	3
L_1, L_2, L_3	50 μH, 80 μH, 100 μH
n_1, n_2, n_3	2
f_s	10 kHz
R_{TE}	10 Ω or 30 Ω
$U_{in1}, U_{in2}, U_{in3}$	50 V to 60 V
U_{nom}	100.66 V (compensated)
U^*_{dc}	100 V
k_1, k_2, k_3	0 to 1
k_{pa}, k_{ia}	2.5, 0.25

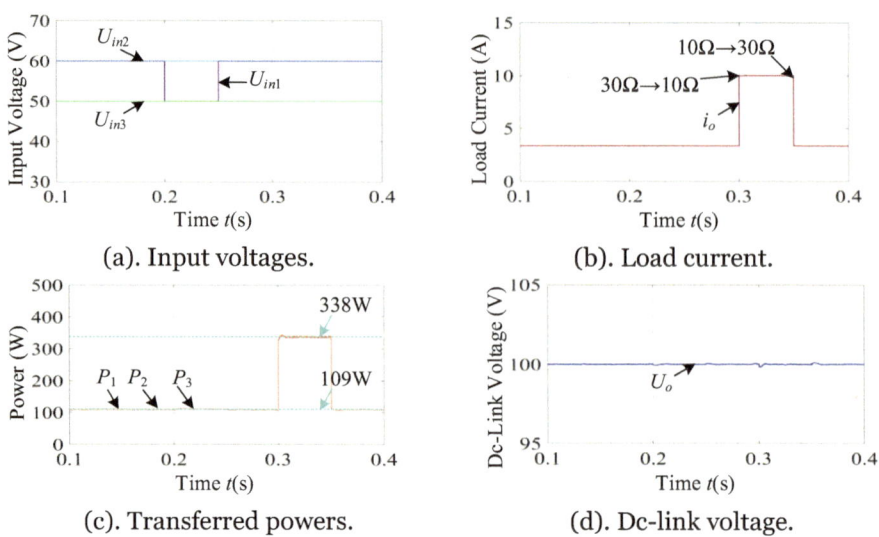

(a). Input voltages.   (b). Load current.

(c). Transferred powers.   (d). Dc-link voltage.

Fig. 3.29 Simulation results when the input voltage and load condition are changed under the proposed communication-free power management strategy

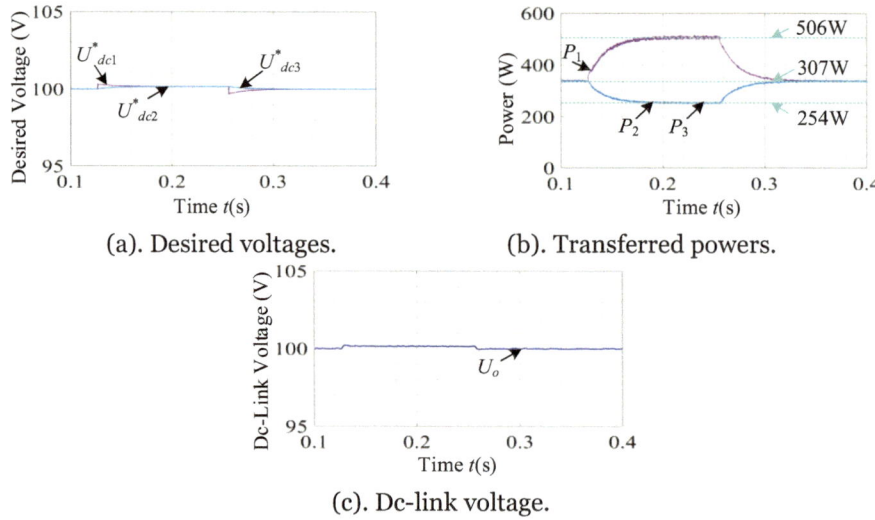

(a). Desired voltages.   (b). Transferred powers.

(c). Dc-link voltage.

Fig. 3.30 Simulation results the droop coefficient k_1 of the first ESU is changed between 0.5 and 1 under the proposed communication-free power management strategy

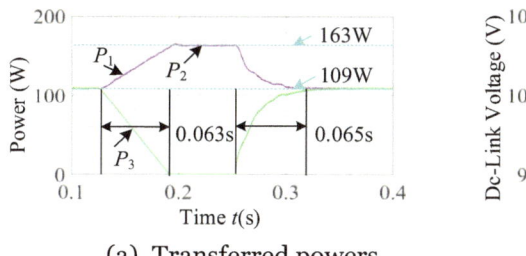
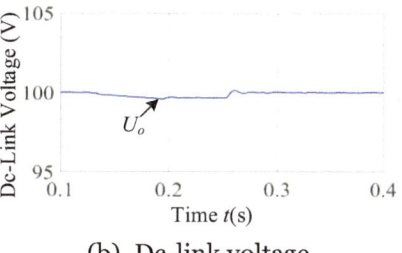

(a). Transferred powers.   (b). Dc-link voltage.

Fig. 3.31 Simulation results when the third ESU is plugged in or plugged out under the proposed communication-free power management strategy

is a little increased, and when the k_1 is decreased, the new steady dc-link voltage is a little reduced. Moreover, as shown in Fig. 3.30b, the power sharing performance of these three ESUs can be strictly determined by the droop coefficients according to (3.39).

Moreover, when the droop coefficients k_1 k_2 and k_3 are 0.5, the load resistor R_{TE} is 30Ω, the input voltage U_{in1} is 60 V, the input voltage U_{in2} is 60 V and the input voltage U_{in3} is 50 V, the simulation results when the third ESU is plugged in or plugged out can be illustrated in Fig. 3.31. As shown in Fig. 3.31a, based on the presented plugging out operation, the transferred power of the third ESU can be reduced gradually, and the other ESUs can compensate the reduced power immediately. When the steady state of the IIOP DAB-based converter system is obtained, the power sharing performance of these three ESUs can be strictly determined by the droop coefficients according to (3.39). Besides, according to Fig. 3.31b, the dc-link voltage can just be affected a little for achieving the steady-state condition again. According to Sect. 3.2.2, when one ESU is plugged out, the new steady dc-link voltage is a little decreased, and when one ESU is plugged in, the new steady dc-link voltage is a little increased. In addition, as shown in Fig. 3.31a, the plug-in and plug-out processes do not take a long time, and the settling time is about 0.06 ms.

Further, when the third DAB is plugged out, the detailed waveforms of the inductance current i_{L3}, the output voltage U_{ab3} of the primary-side H Bridge and the output voltage U_{cd3} of the secondary-side H Bridge of the third DAB converter can be shown in Fig. 3.32. When the switches of the third DAB module are turned off, the inductance current i_{L3} can be consumed quickly.

B. *Experiment Results for the Communication-Free Power Management Strategy.*

Based on the dsPACE MicroLabBox, a small-scale experiment platform is established with two DAB dc-dc converters, and the experiment results are employed to verify the effectiveness of the proposed communication-free power management strategy. The circuit parameters of the IIOP DAB dc-dc converter system are illustrated in Table 3.4. The picture of the experiment platform can be shown in Fig. 3.33.

Fig. 3.32 Detailed waveforms
of i_{L3}, U_{ab3} and U_{cd3} when
the third ESU is plugged out
under the proposed
communication-free power
management strategy

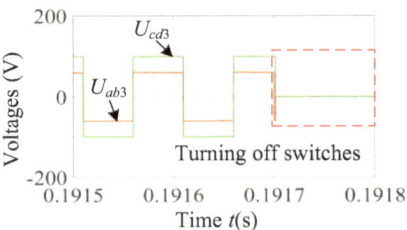

(a). Voltages during plug-out process.

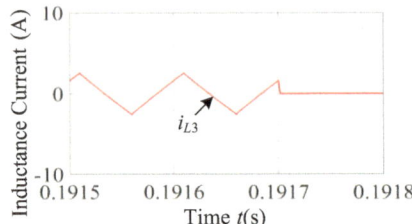

(b). Inductance current during plug-out process.

Table 3.4 Experimental
parameters of the IIOP DAB
dc-dc converter system

Number of DAB modules	2
L_1, L_2	50 μH, 80 μH
n_1, n_2	2
f_s	10 kHz
R_{TE}	16 or 32 Ω
U_{in1}, U_{in2}	30 to 50 V
U_{nom}	61 V (Compensated)
U^*_{dc}	60 V
k_1, k_2	0–1
$k_{p\alpha}, k_{i\alpha}$	2.5, 0.25

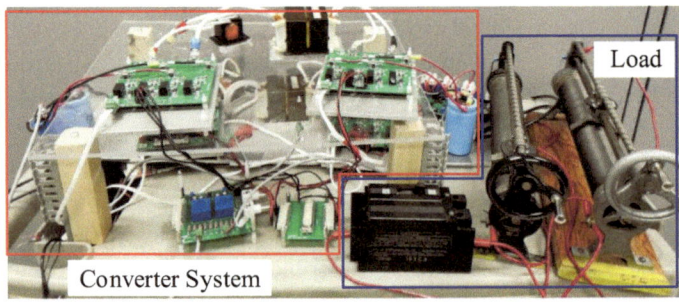

Fig. 3.33 The picture of the experiment platform

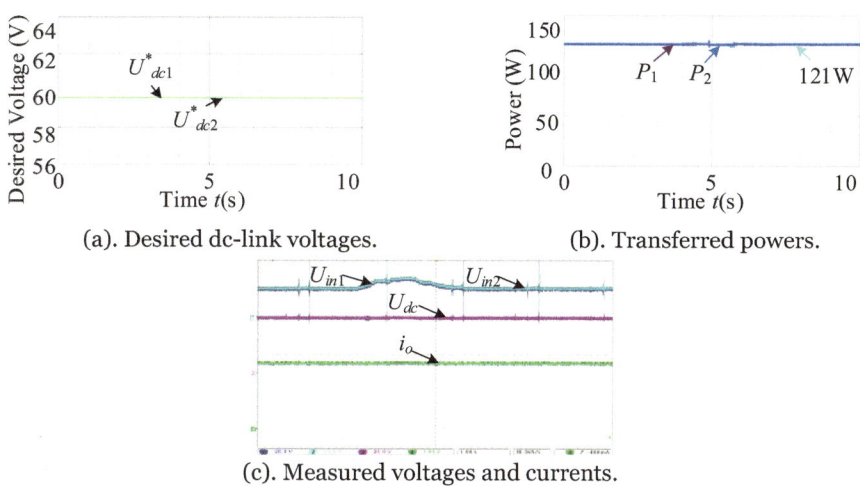

(a). Desired dc-link voltages.(b). Transferred powers.

(c). Measured voltages and currents.

Fig. 3.34 Experiment results when the input voltage is changed under the proposed communication-free power management strategy. (U_{in1} and U_{in2}: 20 V/div; U_{dc}: 20 V/div; i_o: 2 A; t: 1 s/div)

When the droop coefficients k_1 and k_2 are 0.4, the experiment results when the input voltages U_{in1} and U_{in2} are changed between 30 and 40 V can be shown in Fig. 3.34. As shown in Fig. 3.34a, the desired dc-link voltages U^*_{dc1} and U^*_{dc2} of the two DAB modules are not changed, and as shown in Fig. 3.34b, the transferred powers P_1 and P_2 of these two DAB modules are stable during the input-voltage change process. Besides, when the input voltages U_{in1} and U_{in2} are changed between 30 and 40 V, the dc-link voltage U_{dc} can be maintained at its desired value 60 V (see Fig. 3.34c). Therefore, based on the proposed communication-free power management strategy, the excellent dynamic performance can be provided for the IIOP DAB-based ESS when the output voltage of the energy storage equipment is changed.

When the droop coefficients k_1 and k_2 are 0.4, and the input voltages U_{in1} and U_{in2} are 30 V, the experiment results when the load resistor R_{TE} is changed between 16 and 32 Ω can be shown in Fig. 3.35. As shown in Fig. 3.35a, the desired dc-link voltages U^*_{dc1} and U^*_{dc2} of these two DAB modules have a few disturbances when the load resistor R_{TE} are changed, which may be affected by the power loss of the converter system. When the converter system is working at light-load condition, the efficiency is preferred to be low, which means more transferred power should be provided to balance the relationship between the output power and the transferred power. Then, according to (3.39), the desired dc-link voltage can be reduced a little at light-load condition. Moreover, when the load resistor R_{TE} is suddenly changed, the corresponding transferred power can be quickly provided by using the proposed communication-free power management strategy as shown in Fig. 3.35b. Further, as shown in Fig. 3.35c, when the load resistor

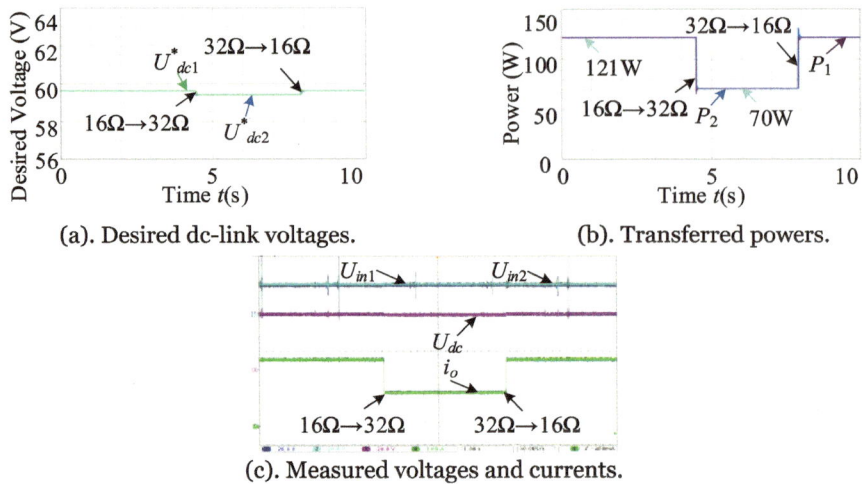

(a). Desired dc-link voltages.   (b). Transferred powers.

(c). Measured voltages and currents.

Fig. 3.35 Experiment results when the load condition is changed under the proposed communication-free power management strategy. (U_{in1} and U_{in2}: 20 V/div; U_{dc}: 20 V/div; i_o: 2 A; t: 1 s/div)

R_{TE} is changed between 16 and 32 Ω, the dc-link voltage can remain at its desired value, and the disturbances of the dc-link voltage can be omitted. Thus, when the load condition is changed, fast responses can be provided for the IIOP DAB-based ESS by using the communication-free power management strategy.

Moreover, when the original droop coefficients k_1 and k_2 are 0.4, the input voltages U_{in1} and U_{in2} are 30 V and the load resistor R_{TE} is 32 Ω, the experiment results when the droop coefficient k_2 of the second DAB module is changed between 0.4 and 0.2 can be illustrated in Fig. 3.36. As shown in Fig. 3.36a, the desired dc-link voltages U^*_{dc1} and U^*_{dc2} of these two DAB modules have a few disturbances when the droop coefficient k_2 are changed. When k_2 is changed, the desired dc-link voltage U^*_{dc2} of the second ESU is changed suddenly. Based on the adjusting function of the droop control structure, the new steady state of the dc-link voltage can be acquired. According to (3.39), when k_2 is increased, the new steady dc-link voltage is a little increased, and when the k_2 is decreased, the new steady dc-link voltage is a little reduced. Moreover, when the droop coefficient k_2 is changed between 0.4 and 0.2, the transferred powers P_1 and P_2 are changed gradually to reach the steady state again as shown in Fig. 3.36b. Further as shown in Fig. 3.36c and d, when the droop coefficient k_2 is changed between 0.2 and 0.4, the dc-link voltage can remain at its desired value, and the disturbances of the dc-link voltage can be omitted. Thus, when the droop coefficient is changed for different power sharing performances, fast responses can be provided for the IIOP DAB-based ESS by using the communication-free power management strategy.

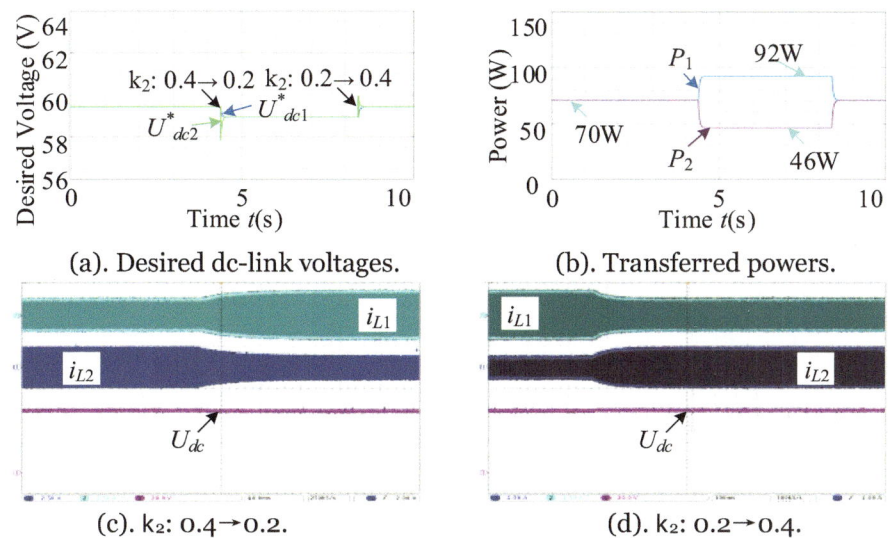

Fig. 3.36 Experiment results when the droop coefficient k_2 of the second ESU is changed between 0.4 and 0.2 under the proposed communication-free power management strategy. (i_{L1} and i_{L2}: 2.5 A/div; U_{dc}: 20 V/div; t: 100 ms/div)

In addition, when the original droop coefficients k_1 and k_2 are 0.4, the input voltages U_{in1} and U_{in2} are 30 V and the load resistor R_{TE} is 32 Ω, the experiment results when the second DAB converter is plugged out and plugged in can be illustrated Fig. 3.37. As shown in Fig. 3.37a, when the second DAB module is plugged out, the desired dc-link voltage U^*_{dc1} is a little reduced, and when the second DAB module is plugged in, the desired dc-link voltage U^*_{dc1} is a little increased since the desired dc-link voltage is affected by the DAB number as demonstrated in (3.49). Moreover, as shown in Fig. 3.37b, the transferred powers P_1 and P_2 are changed gradually to reach the steady state again during the plug-out and plug-in processes. Further as shown in Fig. 3.37c and d, when the second DAB converter is plugged out or plugged in, the dc-link voltage U_{dc} is always close to its expected value 60 V, and the settling time is smaller than 0.1 s. Thus, based on the communication-free power management strategy, the robustness of the dc-link voltage can be ensured when the ESU is plugged out or plugged in, and the plug-in and plug-out processes do not take a long time.

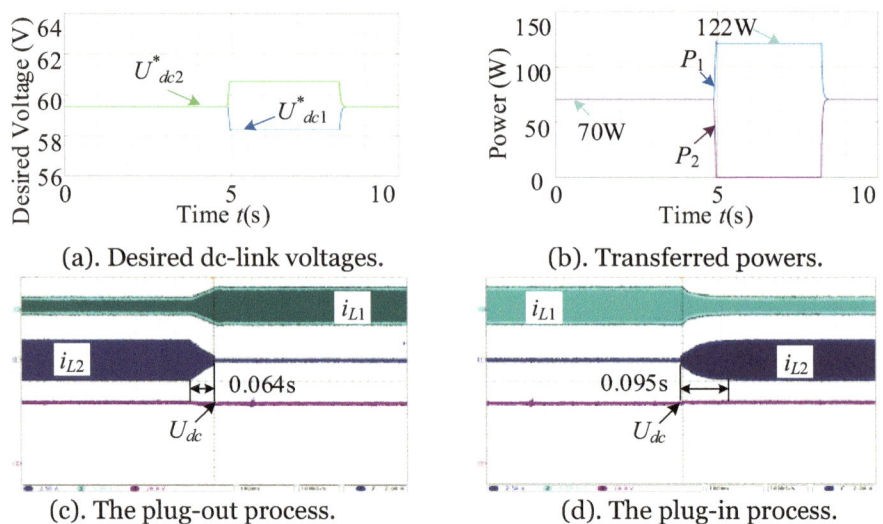

(a). Desired dc-link voltages.   (b). Transferred powers.

(c). The plug-out process.   (d). The plug-in process.

Fig. 3.37 Experiment results when the second ESU is plugged out or plugged in under the proposed communication-free power management strategy. (i_{L1}: 5 A/div; i_{L2}: 2.5 A/div; U_{dc}: 20 V/div; t: 100 ms/div)

3.4   Summary

In this Chapter, a simple tunable power sharing control method is proposed to configure transferred powers for the IPOP DAB dc-dc converter flexibly. Besides, the fast-dynamic performances can be achieved when the input voltage, the load resistor and the power sharing performance are changed. Then, the hot-swap operations are presented for the IPOP DAB dc-dc converter system with only a little influence on the output voltage. Moreover, to guarantee the desired power sharing performance, comprehensive CPE strategies are proposed for facing different conditions. Notably, these CPE methods can also be employed in the IIOP DAB dc-dc converter system with centralized controller. In addition, a communication-free power management strategy is proposed for the IIOP DAB dc-dc converter. Based on the proposed power management strategy, the dc-link voltage can be kept stable when the output voltage of the energy sources, the load condition, and the required power sharing performance of the IIOP DAB converter system are changed. Furthermore, based on the presented plug-in and plug-out operations, the DAB module with new energy storage source can be directly plugged in for increasing the power capacity of the energy storage system without reprogramming operation and influencing the dc-link voltage. Similarly, the energy storage source can also be plugged out without obvious impact on the dc-link voltage.

References

1. N. Hou, Y. W. Li, and H. Tian, "A Reconstructed Circuit Parameters Estimation (RCPE) Strategy of Modular Multiple Dual Active Bridge DC-DC Converters for Power Sharing Control," *2018 IEEE Energy Conversion Congress and Exposition (ECCE)*, 2018, pp. 6715–6720.
2. H. W. Bode, and C. E. Shannon, "A Simplified Derivation of Linear Least Square Smoothing and Prediction Theory," *Proceedings of the IRE,* vol. 38, no. 4, pp. 417–425, 1950.
3. X. Deng, Y. Tang, Y. Zhang, L. Ren, H. Liu, Y. Gong, and J. Li, "An Experimental and Numerical Study on the Inductance Variation of HTS Magnets," *IEEE Transactions on Applied Superconductivity,* vol. 25, no. 3, pp. 1–5, 2015.
4. Y. Lin, Y. Wang, S. Wang, and H. Li, "Sensorless current estimation and sharing in multiphase input-parallel output-parallel DC-DC converters." pp. 1–6.
5. Y. Wang, F. Wang, Y. Lin, and T. Hao, "Sensorless parameter estimation and current-sharing strategy in two-phase and multiphase IPOP DAB DC–DC converters," *IET Power Electronics,* vol. 11, no. 6, pp. 1135–1142, 2018.
6. H. Bai, and C. Mi, "Eliminate Reactive Power and Increase System Efficiency of Isolated Bidirectional Dual-Active-Bridge DC–DC Converters Using Novel Dual-Phase-Shift Control," *IEEE Transactions on Power Electronics,* vol. 23, no. 6, pp. 2905–2914, 2008.
7. B. Zhao, Q. Yu, and W. Sun, "Extended-Phase-Shift Control of Isolated Bidirectional DC–DC Converter for Power Distribution in Microgrid," *IEEE Transactions on Power Electronics,* vol. 27, no. 11, pp. 4667–4680, 2012.
8. F. Krismer, and J. W. Kolar, "Closed Form Solution for Minimum Conduction Loss Modulation of DAB Converters," *IEEE Transactions on Power Electronics,* vol. 27, no. 1, pp. 174–188, 2012.
9. N. Hou, W. Song, Y. Li, Y. Zhu, and Y. Zhu, "A Comprehensive Optimization Control of Dual-Active-Bridge DC–DC Converters Based on Unified-Phase-Shift and Power-Balancing Scheme," *IEEE Transactions on Power Electronics,* vol. 34, no. 1, pp. 826–839, 2019.
10. N. Hou, W. Song, and M. Wu, "Minimum-Current-Stress Scheme of Dual Active Bridge DC–DC Converter With Unified Phase-Shift Control," *IEEE Transactions on Power Electronics,* vol. 31, no. 12, pp. 8552–8561, 2016.
11. Q. Gu, L. Yuan, J. Nie, J. Sun, and Z. Zhao, "Current Stress Minimization of Dual-Active-Bridge DC–DC Converter Within the Whole Operating Range," *IEEE Journal of Emerging and Selected Topics in Power Electronics,* vol. 7, no. 1, pp. 129–142, 2019.
12. G. Oggier, G. O. García, and A. R. Oliva, "Modulation strategy to operate the dual active bridge DC-DC converter under soft switching in the whole operating range," *IEEE Transactions on Power Electronics,* vol. 26, no. 4, pp. 1228–1236, 2011.
13. A. Tong, L. Hang, G. Li, X. Jiang, and S. Gao, "Modeling and Analysis of a Dual-Active-Bridge-Isolated Bidirectional DC/DC Converter to Minimize RMS Current With Whole Operating Range," *IEEE Transactions on Power Electronics,* vol. 33, no. 6, pp. 5302–5316, 2018.
14. N. Hou, and Y. W. Li, "Overview and Comparison of Modulation and Control Strategies for a Nonresonant Single-Phase Dual-Active-Bridge DC–DC Converter," *IEEE Transactions on Power Electronics,* vol. 35, no. 3, pp. 3148–3172, 2020.

The IPOS and ISOP DAB DC-DC Converter Systems

4

In this chapter, a simple tunable power sharing control strategy is proposed for the IPOS DAB dc-dc converter system in Sect. 4.1. Based on this scheme, excellent dynamic performance can also be achieved when the input voltage, the load resistor, and the power sharing performance are changed. Besides, a variant of the proposed scheme is adopted to realize the black-start operation of this DAB system. Moreover, the hot swap operation is presented for the IPOS DAB dc-dc converter with a slight influence on the output voltage. In addition, an input-oriented power sharing control scheme with fast-dynamic response is proposed for the ISOP DAB dc-dc converter system in Sect. 4.2. In addition, an inductance-estimating method is proposed for ensuring the power sharing performance of the ISOP DAB dc-dc converter. Finally, simulation and experiment results are provided to verify the effectiveness of the proposed control schemes for the IPOS DAB dc-dc converter system and the ISOP DAB dc-dc converter system in Sect. 4.3. Then, the logic structure of this Chapter can be summarized in Fig. 4.1.

4.1 A Tunable Power Sharing Control Scheme for IPOS DAB DC-DC Converter System

In this section, based on the SPS modulation method, a tunable power sharing control scheme is proposed for the IPOS DAB dc-dc converter system as shown in Fig. 1.10, which is employed to realize the fast-dynamic performance when the input voltage, the load resistor, and the power sharing performance among DAB modules are changed. Moreover, during the start-up process, a black-start operation is used to charge the output voltage for each DAB module synchronously. In addition, the hot swap operations

N. Hou, *High-Robust Control Schemes for Dual-Active-Bridge-Based DC–DC Converter Systems in Renewable Energy Applications*, Synthesis Lectures on Power Electronics, https://doi.org/10.1007/978-3-031-72963-8_4

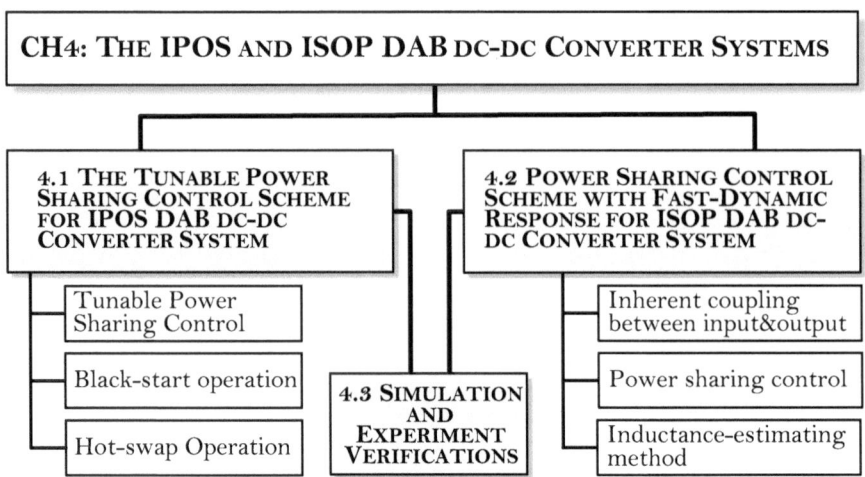

Fig. 4.1 Logic structure of this chapter

are presented to realize the plugging-in or the plugging-out process of the single DAB module.

4.1.1 The Tunable Power Sharing Control Strategy

Traditionally, the inductance of the dc-dc converter such as buck and boost plays an important role in the modeling analysis. Nevertheless, since the transferred current of DAB dc-dc converter can be determined by the circuit parameter and the phase-shift ratio, the middle ac inductance can be neglectful. Thus, the simplified circuit of the IPOS DAB dc-dc converter system can be shown in Fig. 4.2, where the input-side current and the output-side current can be modeled as the controllable current source.

According to Fig. 4.2, the average model of the IPOS DAB dc-dc converter can be expressed as,

$$C_{o\alpha}\frac{dU_{co\alpha}}{dt} = i_{o\alpha} - i_o \, (\alpha \in [1, m]) \tag{4.1}$$

For the IPOS DAB dc-dc converter systems, the power sharing performance is determined by the output-capacitor voltage values $U_{co\alpha}$ of each converter as,

$$P_1:P_2:....:P_m = U_{co1} : U_{co2}... : U_{com} \tag{4.2}$$

Therefore, the power sharing performance of the IPOS DAB dc-dc converter can be realized by adjusting $U_{co\alpha}$. Moreover, the relationship between the change in capacitor

Fig. 4.2 The simplified circuit of IPOS DAB dc-dc converter system

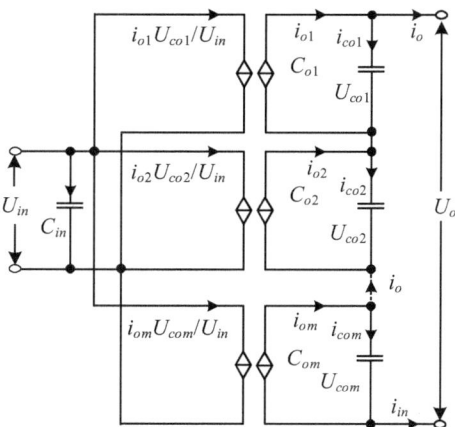

voltage $\Delta U_{co\alpha}$ and the charging current $i_{co\alpha}$ in a switching period T_s can be described as [1],

$$i_{co\alpha} = \frac{\Delta U_{co\alpha} C_{o\alpha}}{T_s} \qquad (4.3)$$

where $C_{o\alpha}$ is the output capacitor for each DAB module. According to Fig. 4.2, the output-side circuit of the IPOS DAB converter system can be shown in Fig. 4.3, where the current flowing between two close DAB converters is equivalent to i_o according to Kirchhoff's Current Law (KCL). Then, the output current of each DAB module $i_{o\alpha}$ can be expressed as,

$$i_{o\alpha} = i_o + i_{co\alpha} \qquad (4.4)$$

Moreover, the transferred current $i_{T\alpha}$ of DAB dc-dc converter based on the SPS modulation method can be shown as [2],

$$i_{T\alpha} == \frac{P_\alpha}{U_{co\alpha}} = \frac{U_{in\alpha} D_\alpha (1 - |D_\alpha|) T_s}{2 n_\alpha L_\alpha} \qquad (4.5)$$

here P_α is the transferred power, $U_{in\alpha}$ is the input voltage, D_α is the phase-shift ratio, n_α is the transformer turn ratio and L_α is the inductance for each DAB dc-dc module. D_α can be calculated by $i_{T\alpha}$ as,

$$\begin{cases} D_\alpha = \frac{1}{2} - \sqrt{\frac{1}{4} - \frac{2 n_\alpha L_\alpha i_{T\alpha}}{U_{in\alpha} T_s}} & (i_{T\alpha} \geq 0) \\ D_\alpha = -\frac{1}{2} + \sqrt{\frac{1}{4} + \frac{2 n_\alpha L_\alpha i_{T\alpha}}{U_{in\alpha} T_s}} & (i_{T\alpha} < 0) \end{cases} \qquad (4.6)$$

Ignoring power losses, the transferred power is equivalent to the output power of each DAB module, and $i_{T\alpha}$ can be equivalent to $i_{o\alpha}$. However, the power losses of the IPOS

Fig. 4.3 The output-side
circuit of the IPOS DAB dc-dc
converter system

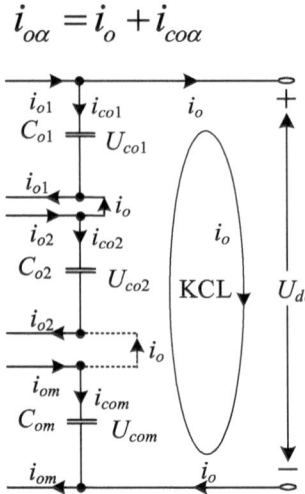

$$i_{o\alpha} = i_o + i_{co\alpha}$$

DAB dc-dc converter in actual system cannot be neglected. Therefore, a compensation coefficient k_C should be introduced to eliminate the difference between $i_{o\alpha}$ and $i_{T\alpha}$ as,

$$i_{T\alpha} = k_C i_{o\alpha} \tag{4.7}$$

In terms of accurately maintaining output dc-link voltage of the IPOS DAB dc-dc converter, k_C can be calculated by a virtual output voltage U_{dcv} generated by the outer-voltage loop integral controller with the control input $(U^*_{dc}-U_{dc})$. Then, k_C can be expressed as,

$$k_C = \frac{U_{dcv}}{U_{dc}} \tag{4.8}$$

According to (4.7), (4.4) can be further expressed as,

$$i_{T\alpha} = \frac{U_{dcv}}{U_{dc}} i_o + \frac{U_{dcv}}{U_{dc}} i_{co\alpha} \tag{4.9}$$

As shown in (4.9), $i_{T\alpha}$ can be divided into two parts for meeting the output-current requirement for load and adjusting capacitor voltage for each DAB dc-dc converter as,

$$\begin{cases} i'_o = \dfrac{U_{dcv}}{U_{dc}} i_o \\ i'_{co\alpha} = \dfrac{U_{dcv}}{U_{dc}} i_{co\alpha} \approx i_{co\alpha} \end{cases} \tag{4.10}$$

In (4.10), when desired output dc-link voltage of the IPOS DAB dc-dc converter is achieved, U_{dcv} is approximately equal to U_{dc}. Therefore, in terms of adjusting the capacitor voltage for each DAB converter, $i'_{co\alpha}$ can be seen as $i_{co\alpha}$. Besides, combining (4.3)

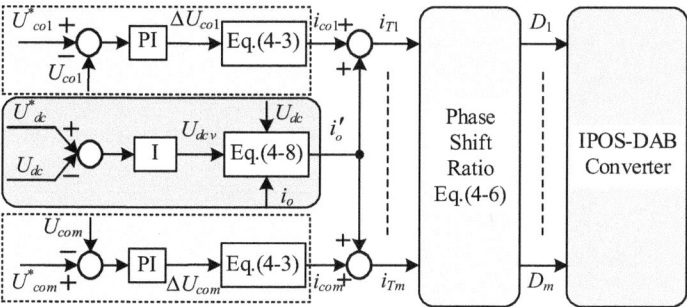

Fig. 4.4 The block schematic of the tunable power sharing control strategy

and (4.10), the block schematic of the tunable power sharing control strategy can be shown in Fig. 4.4, and the tunable power sharing control strategy for the IPOS DAB dc-dc converter system can be implemented. The I controller is employed to compensate the error between U_{dc} and U^*_{dc}, and the PI controllers with the identical integral and proportionality parameters are used to adjust $U_{co\alpha}$ synchronously. At the beginning of each switching cycle, $U_{in\alpha}$, $U_{co\alpha}$ and i_o are measured, and U_{dc} can be calculated as the sum of $U_{co\alpha}$. Through the PI controller, $\Delta U_{co\alpha}$ can be obtained by $U_{co\alpha}$ and $U^*_{co\alpha}$, and then, $i_{co\alpha}$ can be calculated according to (4.3). Similarly, from the I controller, U_{dcv} can be obtained by U_{dc} and U^*_{dc}, and i'_o can be acquired by using (4.10). Finally, according to (4.6), D_α for each DAB module can be obtained. Generally, in the battery- or supercapacitor-based ESS system, $U^*_{co\alpha}$ can be acquired for implementing the state-of-charge-balancing control of batteries and super capacitors [3–5].

4.1.2 Black-Start Operation and Hot-Swap Processes

A. *Black Start of the IPOS DAB dc-dc Converter System.*

Under the tunable power sharing control strategy, fluctuations of capacitor voltages are difficult to avoid, since the change of i'_o cannot be consistent with the change of i_o during the start-up process, and extra power should be employed to charge capacitor voltages. To realize black start of the IPOS DAB dc-dc converter, a variant tunable power sharing control strategy should be used, which can be shown in Fig. 4.5.

Different from the original tunable power sharing control strategy, i_o is employed as a current feedforward control, and the load current can be satisfied continuously during the start-up process. Therefore, the change of $U_{co\alpha}$ can be synchronized. When U^*_{dc} is reached, U_{dcv} in tunable power sharing control scheme can be calculated as,

$$U_{dcv} = U_{dc} \tag{4.11}$$

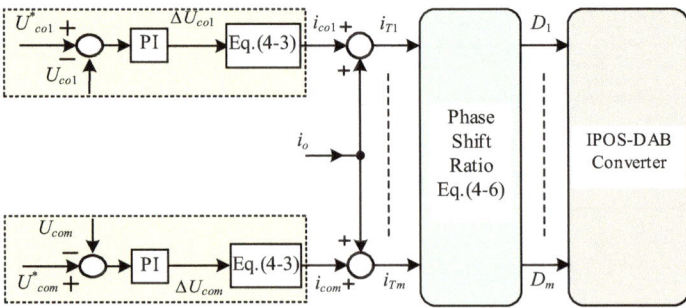

Fig. 4.5 The block schematic of the variant tunable power sharing control strategy for black start

Based on (4.11), the variant tunable power sharing control scheme can be switched into tunable power sharing control scheme smoothly.

B. *Hot-Swap Process of the One DAB Module.*

To repair or replace energy storage equipment such as the battery and the super capacitor, the soft plug-in and plug-out control of DAB module is crucial for the IPOS DAB dc-dc converter system. When the βth ESS equipment is required to be plugged out, the process is divided into two steps for plugging out this source: (1) controlling the corresponding capacitor voltage towards zero, (2) shorting out the corresponding second-side H Bridge of the corresponding DAB module. During the first step, the capacitor voltage $U_{co\beta}$ should be reduced gradually based on the tunable power sharing control scheme. When $U_{co\beta}$ is smaller than the limited value U_{LM}, the switches of the second-side H Bridge can be turned on, and U_{LM} can be expressed as,

$$U_{LM} < i_{LM} R_{on} \tag{4.12}$$

where i_{LM} is the current rating of employed switches and R_{on} is the corresponding conducting resistor. This DAB module can be bypassed as shown in Fig. 4.6, and the corresponding energy storage equipment can be taken down.

In addition, when a new energy storage equipment is to be plugged in the IPOS DAB dc-dc system again, the plug-in process can be implemented by using tunable power sharing control strategy with the new power sharing ratio for each DAB dc-dc module. Then, when the desired power sharing performance is achieved, this energy storage module can be plugged in completely.

Fig. 4.6 The current flowing condition when one DAB module is bypassed

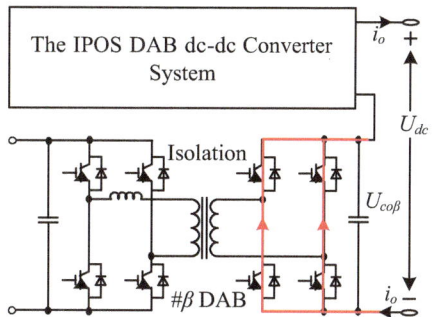

C. Additional Applications of the Tunable Power Sharing Control Scheme.

The basic operating principle of the proposed tunable power sharing control strategy is the accurate compensation of the load current and the precise adjustment of the capacitor voltages. Therefore, the relationship of the transferred current and the control value of the dc-dc converter module should be determined, such as in (4.5) for the DAB dc-dc converter. According to this principle, the proposed method can be employed for the output-series dc-dc converter system with an ac-inductance-based converter module. For the ac-inductance-based converter such as half-bridge dc-dc converter [6], full bridge dc-dc converter [7] and three-phase DAB dc-dc converter [8], the ac voltages are generated on both sides of its inductance resulting in ac inductance current. Besides, the transferred power and the transferred current of this kind of converter can be determined by the control value in each switching period directly [9, 10]. For example, the transferred current i_{TF} of full bridge dc-dc converter system can be expressed as,

$$i_{TF} = \begin{cases} \dfrac{(nU_{inF} - U_{oF})U_{inF}d^2}{4n_F f_F L_F U_{oF}} & \left(0 < \varphi \le \dfrac{U_{oF}}{nU_{inF}}\right) \\[4mm] \dfrac{U_{inF}}{8n_F f_F L_F}d(2-d) - \dfrac{U_{oF}^2}{8n_F^3 U_{inF} f_F L_F} & \left(\dfrac{U_{oF}}{nU_{inF}} < \varphi \le 1\right) \end{cases} \tag{4.13}$$

here U_{inF} is input voltage, n_F is the transformer turn ratio, f_F is the switching frequency, L_F is the inductance, U_{oF} is the output voltage, and d is the phase-shift ratio for the full bridge dc-dc converter system. Each i_{TF} can be determined by a certain phase-shift ratio d. Based on (4.13), d can be calculated by i_{TF} and the tunable power sharing control strategy can be also implemented.

4.2 A Power Sharing Control Scheme with Fast-Dynamic Response for ISOP DAB DC-DC Converter System

In this section, based on the minimum-current-stress phase-shift modulation method, a power sharing control scheme with fast-dynamic response is proposed for the ISOP DAB dc-dc converter system as shown in Fig. 1.11, which can provide the fast-dynamic performance when the input voltage, the load resistor and the power sharing performance among DAB modules are changed. Moreover, to ensure the desired power sharing performance of the ISOP DAB dc-dc converter system, the general inductance-estimating method is proposed.

4.2.1 The Inherent Coupling Phenomenon Between Regulations of Input Voltages and Output Voltage

In this section, the average model of the ISOP DAB dc-dc converter system is presented. Besides, the current distributions on both the primary side and the secondary side are demonstrated, which can be employed to analyze the coupling relationship between the adjustment of the input voltages and the output voltage. Traditionally, the inductance of the dc-dc converter such as buck and boost plays an important role in the modeling analysis. Nevertheless, since the transferred current of DAB dc-dc converter can be determined by the circuit parameter and the phase-shift ratio, the middle ac inductance can be neglectful [9, 11]. Thus, the simplified circuit of the ISOP DAB dc-dc converter system can be shown in Fig. 4.7, where the input-side current and the output-side current can be modeled as the controllable current source.

According to Fig. 4.7, the average model of the ISOP DAB dc-dc converter can be expressed as,

Fig. 4.7 The simplified circuit of ISOP DAB dc-dc converter system

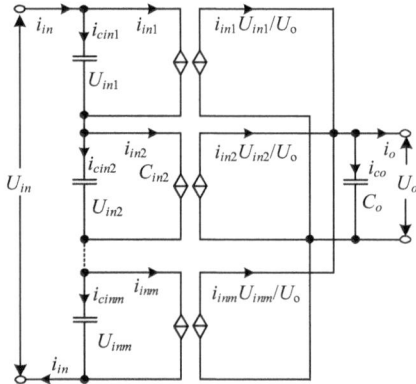

$$\begin{cases} C_{in\alpha} \dfrac{dU_{in\alpha}}{dt} = i_{in} - i_{in\alpha} \\ C_o \dfrac{dU_o}{dt} = \displaystyle\sum_{\alpha=1}^{m} \dfrac{i_{in\alpha} U_{in\alpha}}{U_o} - i_o \end{cases} \quad (\alpha \in [1, m]) \tag{4.14}$$

According to (4.14), since the total disturbances of the input voltage for each DAB module should be zero, the input current i_{in} can be expressed as,

$$i_{in} = \frac{\sum_{\alpha=1}^{m} \frac{i_{in\alpha}}{C_{in\alpha}}}{\sum_{\alpha=1}^{m} \frac{1}{C_{in\alpha}}} = \frac{\sum_{\alpha=1}^{m} i_{in\alpha}}{m} \bigg|_{\text{same input capacitors}} \tag{4.15}$$

In (4.15), when the ISOP DAB dc-dc converter is at steady state condition, the transferred current for each DAB module $i_{in\alpha}$ should be equivalent to the input current i_{in}. Besides, assuming the input capacitors are the same when a variation $\Delta i_{in\beta}$ is added to the βth DAB module for adjusting its input voltage, the input voltage i_{in} can be expressed as,

$$i_{in} = \frac{\displaystyle\sum_{\alpha=1}^{m} i_{in\alpha} + \Delta i_{in\beta}}{m} = i'_{in} + \frac{\Delta i_{in\beta}}{m} \tag{4.16}$$

where i'_{in} is the required input current for the load side at steady-state condition. The capacitor charging current $i_{cin\alpha}$ for each DAB module can be expressed as,

$$\begin{cases} i_{cin\beta} = -\dfrac{m-1}{m} \Delta i_{cin\beta} \\ i_{cin\alpha} = \dfrac{1}{m} \Delta i_{cin\beta} \; (\alpha \neq \beta) \end{cases} \tag{4.17}$$

On this condition, the transferred current to the output side can be calculated as,

$$\sum_{\alpha=1}^{m} \frac{i_{in\alpha} U_{in\alpha}}{U_o} = \sum_{\alpha=1}^{m} \frac{i'_{in} U_{in\alpha}}{U_o} + \frac{\Delta i_{in\beta} U_{in\beta}}{U_o} = U_o i_o + \frac{\Delta i_{in\beta} U_{in\beta}}{U_o} \tag{4.18}$$

According to (4.18), the charging current of the output capacitor i_{co} can be calculated as,

$$i_{co} = \frac{\Delta i_{in\beta} U_{in\beta}}{U_o} \tag{4.19}$$

The output voltage cannot be stable since the charging current of the output capacitor is not zero. Moreover, when all the input voltage $U_{in\alpha}$ should be adjusted positively for the desired power sharing performance, the variation $\Delta i_{in\alpha}$ for each DAB module can be expressed as,

$$\Delta i_{in\alpha} = C_\alpha \Delta U_{in\alpha} \left(\sum_{\alpha=1}^{m} \Delta U_{in\alpha} = 0 \right) \qquad (4.20)$$

Assuming the input capacitors are the same, the sum of the transferred current variations can be expressed as,

$$\sum_{\alpha=1}^{m} \Delta i_{in\alpha} = 0 \qquad (4.21)$$

According to (4.15) and (4.21), the input current i_{in} cannot be changed. Then, combining Fig. 4.7 and (4.18), the charging current i_{co} of the output capacitor can be expressed as,

$$i_{co} = \sum_{\alpha=1}^{m} \frac{\Delta i_{in\alpha} U_{in\alpha}}{U_o} \neq 0 \quad \left(\sum_{\alpha=1}^{m} \Delta i_{in\alpha} = 0 \right) \qquad (4.22)$$

During the transient process for adjusting the input voltages, the input voltages $U_{in\alpha}$ for each DAB module are usually not the same, so the charging current for the output capacitor is not zero. Thus, the disturbance of the output voltage is not evitable. In reverse, if i_{co} is forced to zero in (4.22), the disturbance of the output voltage can be omitted, but the sum of the transferred current variations cannot be zero. Therefore, the change of the input voltages $U_{in\alpha}$ cannot be the same as the requirement. In terms of adjusting the input voltages for each DAB module, the coupling of the regulation of the input voltages and the adjustment of the output voltage cannot be reduced.

4.2.2   The Proposed Power Sharing Control Scheme with Fast-Dynamic Response

In this section, the novel power sharing control scheme with fast-dynamic response is proposed for the ISOP DAB dc-dc converter. To reduce the power loss, an existing minimum-current-stress modulation method is adopted [12]. Then, the proposed power sharing control scheme is presented and analyzed, which can also provide fast-dynamic response when the input voltage and load resistor are changed.

A. *Minimum-Current-Stress Modulation Method.*

The waveforms of the minimum-current-stress modulation method can be shown in Fig. 4.8, where $D_{\alpha 1} \sim D_{\alpha 3}$ are the phase-shift ratios. Based on this modulation method, the soft switching performance can be obtained during the whole power range, and the minimum conduction power loss can also be achieved. Then, the corresponding phase-shift

Fig. 4.8 The minimum-current-stress modulation method under different voltage conditions

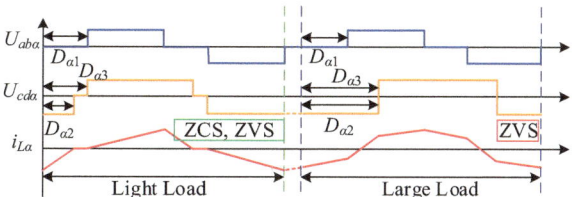

(a). Phase-shift modulation methods when k>1.

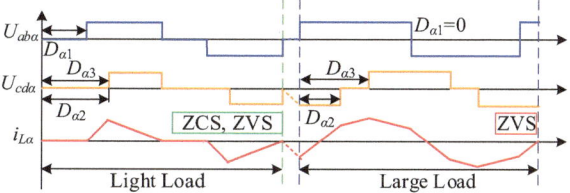

(b). Phase-shift modulation methods when k<1.

ratios $D_{\alpha 1} \sim D_{\alpha 3}$ can be obtained from Table 4.1, where $I_{T\alpha}$ is the transferred current, n_α is the transformer turn ratio, L is the middle inductance of the αth DAB dc-dc module.

According to Fig. 4.7, the transferred current $I_{T\alpha}$ for each DAB module can be expressed as,

$$I_{T\alpha} = \frac{i_{in\alpha} U_{in\alpha}}{U_o} \quad (4.23)$$

Moreover, based on Table 4.1, the current-level modulation operation can be realized, and the corresponding schematic can be shown in Fig. 4.9.

As shown in Fig. 4.9, if the transferred current $I_{T\alpha}$ is employed as the effective control value for DAB dc-dc converter, the current-level modulation can be employed. Based on some existing modulation methods [12], the phase-shift ratios can be determined by the transferred current or power even during transient process. Thus, when the DAB converter is modeled, the underlying phase-shift ratios can be avoided, which can significantly simplify the modeling analysis about the DAB converter.

B. *The Input-Oriented Power Sharing Control Method with Fast-Dynamic Response.*

According to (4.22), when the input voltages should be controlled positively, the influence on the output voltage is unavoidable. So, to reduce this influence, the changes $\Delta U_{in\alpha}$ of each input voltage should be relatively small. Then, the charging current $i_{cin\alpha}$ for each input-side capacitor can be calculated as,

$$i_{cin\alpha} = \frac{\Delta U_{in\alpha} C_{in\alpha}}{T_s} \quad (4.24)$$

Table 4.1 Optimized solutions of minimum-current-stress modulation method with transferred current

Voltage conditions	Unified transferred current	Range of $i_{T\alpha}$	Middle variable	Phase-shift ratio
$k > 1$	$i_{T\alpha} = \dfrac{8L_\alpha I_{T\alpha}}{n_\alpha U_{in\alpha} T_s}$	$0 \le i_{T\alpha} < 2\dfrac{k_\alpha - 1}{k_\alpha^2}$	$D_{\alpha 1} = 1 - \sqrt{\dfrac{i_{T\alpha}}{2(k_\alpha - 1)}}$	$\begin{cases} D_{\alpha 2} = (k_\alpha - 1)(1 - D_{\alpha 1}) \\ D_{\alpha 3} = D_{\alpha 1} \end{cases}$
		$2\dfrac{k_\alpha - 1}{k_\alpha^2} \le i_{T\alpha} \le 1$	$D_{\alpha 1} = (k_\alpha - 1)\sqrt{\dfrac{1 - i_{T\alpha}}{k_\alpha^2 - 2k_\alpha + 2}}$	$\begin{cases} D_{\alpha 2} = \dfrac{k_\alpha - 2}{2(k_\alpha - 1)} D_{\alpha 1} + \dfrac{1}{2} \\ D_{\alpha 3} = \dfrac{k_\alpha - 2}{2(k_\alpha - 1)} D_{\alpha 1} + \dfrac{1}{2} \end{cases}$
$k \le 1$		$0 \le i_{T\alpha} < 2(k_\alpha - k_\alpha^2)$	$D_{\alpha 1} = 1 - \sqrt{\dfrac{i_{T\alpha}}{2k_\alpha(1 - k_\alpha)}}$	$\begin{cases} D_{\alpha 2} = 0 \\ D_{\alpha 3} = k_\alpha D_{\alpha 1} - k_\alpha + 1 \end{cases}$
		$2(k_\alpha - k_\alpha^2) \le i_{T\alpha} \le 1$	$D_{\alpha 2} = \dfrac{1}{2}\left(1 - \sqrt{\dfrac{1 - i_{T\alpha}}{2k_\alpha^2 - 2k_\alpha + 1}}\right)$	$\begin{cases} D_{\alpha 1} = 0 \\ D_{\alpha 3} = 2k_\alpha D_{\alpha 2} - D_{\alpha 2} - k_\alpha + 1 \end{cases}$

Fig. 4.9 The schematic of current-level modulation for DAB converter

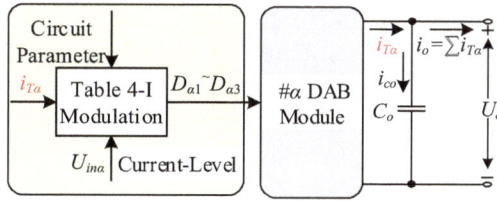

Moreover, to deal with the disturbance of the load condition, the required output current for the ISOP DAB dc-dc converter can be calculated as,

$$i_o^* = \frac{U_o^*}{R_{eq}} = \frac{i_o U_o^*}{U_o}$$

(4.25)

where U_o^* is the desired output voltage and R_{eq} is the equivalent load resistor. Since the power losses cannot be neglected in the actual converter system, a compensation value k_c should be introduced by using the fuzzy adjustment of the PI controller. The compensation value k_c can be the output value of the PI controller with the output voltage and its desired value as inputs. Besides, the static error can be reduced, and (4.25) can be furtherly expressed as,

$$i_o^* = k_c \frac{i_o U_o^*}{U_o}$$

(4.26)

In (4.26), the all DAB modules should share this required output current i_o^*. Traditionally, the existing strategies prefer to divide this required output current evenly for each dc-dc converter module [13–17]. However, this operation usually influences on the regulation of the input voltages since the DAB modules should have different transferred current abilities with different input voltages, and the input-side capacitors should compensate the transferred currents. Thus, the adjustment of the output voltage influences the input voltages when the input voltages are not the same. To meet the current transferred abilities of each DAB module, the transferred current i_α for meeting the requirement of output side should be expressed as,

$$i_\alpha = \frac{i_o^* U_{in\alpha}}{U_{in}}$$

(4.27)

According to (4.27), when the input voltage and the load condition is changed, the influence on the voltage sharing performance on the input side can be significantly eliminated. Besides, combining Table 4.1, (4.3), (4.26) and (4.27), the proposed input-oriented power sharing control scheme with fast-dynamic response can be demonstrated in Fig. 4.10. At the beginning of the switching period, the input voltages, the output voltage, and the load current are measured. Based on the error of the output voltage and its desired value, the compensated value k_c can be obtained from the PI controller. Moreover, according to (4.26), the required output current can be calculated, and based on (4.27), the required transferred current i_α of each DAB module for meeting the requirement of load side can be obtained. In addition, based on the error of the input voltages and their desired values, the required change values for each input voltage can be obtained based on a P controller. Then, according to (4.3) and $U_{in\alpha}/U_o$, the required transferred current Δi_α for adjusting the input voltages can be obtained. However, since the input voltages of each module may be very different, the item $U_{in\alpha}/U_o$ should usually result in large difference of transferred currents for charging and discharging the input-side capacitors. So, it is better to delete this item in the control system. Furthermore, the total required transferred current $I_{T\alpha}$ can be calculated as the difference of i_α and Δi_α. Finally, based on Table 4.1, the corresponding phase-shift ratios can be obtained, and the proposed strategy can be implemented for positively adjusting the input voltages and dealing with the disturbance of the total input voltage and the load condition.

C. The Designs of the P Parameter and the PI Parameters in the Proposed Scheme.

Combining (4.20), (4.22) and Fig. 4.10, when the input voltages of DAB modules are adjusted, the disturbance of the output voltage ΔU_o in a switching period can be expressed as,

$$\Delta U_o = \sum_{\alpha=1}^{m} \frac{\Delta i_{in\alpha} U_{in\alpha} T_s}{C_o U_o} = \frac{\sum\limits_{\alpha=1}^{m} k_{pin}(U_{in\alpha}^* - U_{in\alpha})C_\alpha U_{in\alpha} T_s}{C_o U_o} \tag{4.28}$$

where k_{pin} is the proportional parameter of the P controller. Assuming the allowable output-voltage disturbance ΔU_{omax}, k_{pin} can be calculated as,

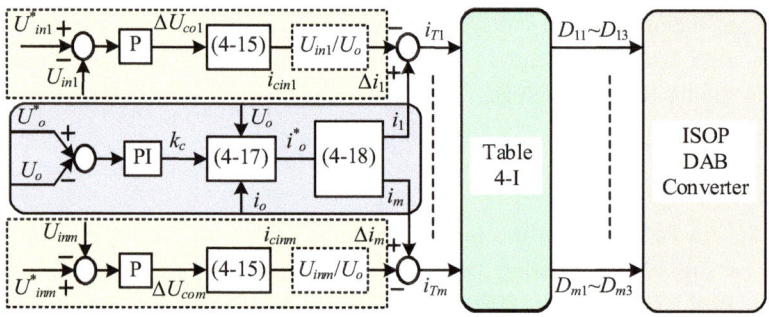

Fig. 4.10 The diagram of the proposed input-oriented power sharing control scheme with fast dynamic response for ISOP DAB dc-dc converter

Fig. 4.11 The simplified circuit of the ISOP DAB dc-dc converter

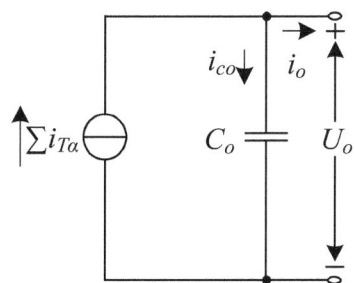

$$k_{pin} \leq \frac{C_o U_o \Delta U_o}{\sum\limits_{\alpha=1}^{m} (U_{in\alpha}^* - U_{in\alpha}) C_\alpha U_{in\alpha} T_s} \tag{4.29}$$

Moreover, since storage energy in middle inductance at the beginning and the end of a switching period can be regarded as the same at the steady-state condition and during the transient process, the simplified circuit of the ISOP DAB dc-dc converter can be shown in Fig. 4.11.

To reduce the influence on the output voltage during the adjustment of the input voltages, the variations of the input-capacitor currents should be very small. Thus, these branches for adjusting the input voltages can be eliminated when analyzing the PI parameters of the middle PI controller in Fig. 4.10. Then, the control schematic for regulating the output voltage can be demonstrated in Fig. 4.12.

According to Fig. 4.12, the transfer function with control-loop delay can be expressed as,

$$H(s) = \frac{U_o^* i_o}{U_o} \frac{k_p s + k_i}{s} \frac{1}{sC_o} e^{-sT_s} \tag{4.30}$$

In addition, the main circuit parameters of the ISOP DAB dc-dc converter can be shown in Table 4.2. Combining (4.30), the bode diagram can be demonstrated as Fig. 4.13.

As shown in Fig. 4.13, the phase margin at cross-over frequency is bigger than 45° as 90°, so the stability of the proposed power sharing control scheme with fast-dynamic response can be ensured for the ISOP DAB dc-dc converter.

Fig. 4.12 The control schematic for regulating the output voltage

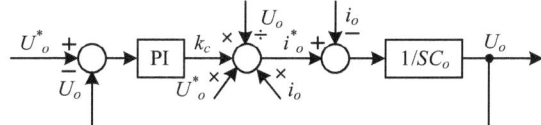

Table 4.2 Circuit parameters of the ISOP DAB dc-dc converter system

Parameter	Value
L_1, L_2	40 μH
n_1, n_2	1
f_s	40 kHz
R	15~50 Ω
U_{in}	100~120 V
U^*_o	50 V
k_p, k_i, k_{Pin}	0.05, 0.005, 0.2
C_{in1}, C_{in2}, C_o	1 mF

Fig. 4.13 The bode diagram of the control loop for regulating the output voltage

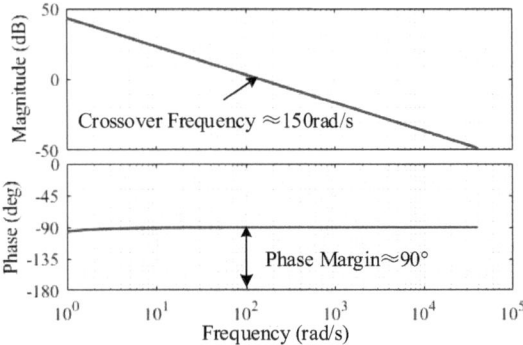

4.2.3 The Inductance-Estimating Method for Ensuring the Desired Power Sharing Performance

According to Table 4.1, the circuit parameters are employed to realize the power sharing performance of the ISOP DAB dc-dc converter system. So, if the employed inductance is not accurate, the power sharing performance should be affected without the integral function for adjusting the input voltages. When the steady-state condition is achieved, the relationship between the input power and the output power for each DAB module can be expressed as,

$$i_{in}U_{in1} : ...i_{in}U_{in\alpha} : ...i_{in}U_{inm} = i'_{T1}U_o : ...i'_{T\alpha}U_o : ...i'_{Tm}U_o \qquad (4.31)$$

where $i'_{T\alpha}$ is the actual transferred power for each DAB module. Assuming the actual inductance for each DAB module is L'_α, (4.31) can be further expressed as,

$$U_{in1} : ...U_{in\alpha} : ...U_{inm} = \frac{i_{T1}L_1}{L'_1} : ...\frac{i_{T\alpha}L_\alpha}{L'_\alpha} : \frac{i_{Tm}L_m}{L'_m} \qquad (4.32)$$

By using the first inductance L'_1 as reference, the other inductance can be calculated as,

$$L'_\alpha = \frac{i_{T\alpha} L_\alpha L'_1 U_{in1}}{i_{T1} L_1 U_{in\alpha}} \tag{4.33}$$

According to (4.33), the total transferred current i_T can be calculated as,

$$i_T = \frac{i_{T1} L_1}{L'_1} + \sum_{\alpha=2}^{m} \frac{i_{T1} L_1 U_{in\alpha}}{L'_1 U_{in1}} \tag{4.34}$$

Based on Energy Conversion Law, the total transferred current i_T should be equivalent to the desired output current i^*_o as,

$$i_T = i^*_o = \frac{i_{T1} L_1}{L'_1} + \sum_{\alpha=2}^{m} \frac{i_{T1} L_1 U_{in\alpha}}{L'_1 U_{in1}} \tag{4.35}$$

According to (4.35), the actual inductance L'_1 of the first DAB module can be expressed as,

$$L'_1 = \frac{i_{T1} L_1}{i^*_o} + \sum_{\alpha=2}^{m} \frac{i_{T1} L_1 U_{in\alpha}}{i^*_o U_{in1}} \tag{4.36}$$

Combining (4.33), the inductances of other DAB modules can be estimated. Moreover, the transferred current of each DAB module can be expressed as,

$$i_{T\alpha} = \frac{U_{inX} f(D_{\alpha 1}, D_{\alpha 2}, D_{\alpha 3}) T_s}{4 n_\alpha L_\alpha} \tag{4.37}$$

Combining (4.27) and (4.37), the estimated inductance for each DAB module can be expressed as,

$$L'_X = \frac{f(D_{\alpha 1}, D_{\alpha 2}, D_{X3}) U_{in} T_s}{4 i^*_o n_\alpha} \tag{4.38}$$

According to (4.38), even without knowledge of inductance values, the proposed inductance-estimating method can be employed in other control schemes with measurement of load current for the ISOP DAB dc-dc converter system. Moreover, since the inductance-estimating method should be employed in the steady-state condition for more accurate estimated value, the decoupling between the estimating process and the transient process should be realized, especially for dealing with the load-resistor disturbance. In addition, the change of the load current can be employed to stop the estimating process, and the change of the input voltage should be observed for ensuring the steady-state condition of input voltages. Furthermore, when the change of the input voltage of two switching periods is smaller than the peak value of the measurement noise, the estimating operation can be activated again. The time duration of these two switching periods should be big enough such as more than half of the change time of the input voltages.

4.3   Verification

In this section, based on the small-scale experiment platforms, the experiment results are employed to verify the effectiveness of the proposed schemes for the IPOS DAB dc-dc converter system and the ISOP DAB dc-dc converter system. Moreover, for the ISOP DAB dc-dc converter system, the simulation results are also provided for monitoring some middle control values of the proposed power sharing control scheme with fast-dynamic response.

4.3.1   A Tunable Power Sharing Control Scheme for the IPOS DAB DC-DC Converter System

An experiment platform for the IPOS DAB dc-dc converter system with two DAB modules is established to compare with the voltage PI-based (VPI) method and the tunable power sharing control scheme, and the main circuit parameters of the IPOS DAB dc-dc converter are illustrated in Table 4.3. The corresponding experimental platform can be shown in Fig. 4.14.

When R is selected as 56 Ω, Fig. 4.15 shows the start-up process of the IPOS DAB dc-dc converter system under different strategies. As shown in Fig. 4.15, tunable power sharing control strategy can reach U^*_{dc} in a short time as 100 ms, compared with VPI method (200 ms) and variant tunable power sharing control scheme (200 ms). Moreover, based on the variant tunable power sharing control scheme, U_{dc}, U_{co1} and U_{co2} can

Table 4.3 Circuit parameters of the IPOS DAB dc-dc converter system

Parameter	Value
U_{in1}, U_{in2}	30 V
n_1, n_2	1/2
L_1	400 μH
L_2	200 μH
T_s	0.1 ms
R	40 or 56 Ω
C_{o1}	1.0 mF,
C_{o2}	0.5 mF
U^*_o	60 V
f_s	10 kHz

Fig. 4.14 The small-scale experimental platform for IPOS DAB dc-dc converter system with two modules

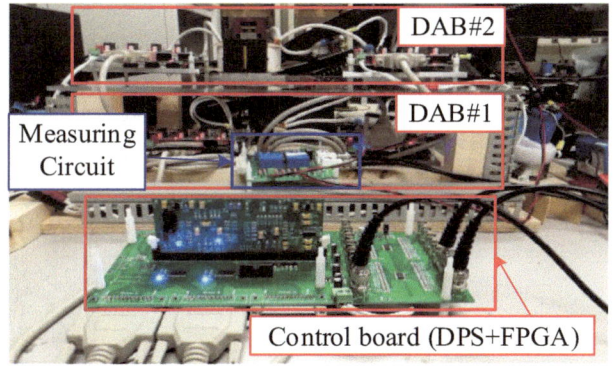

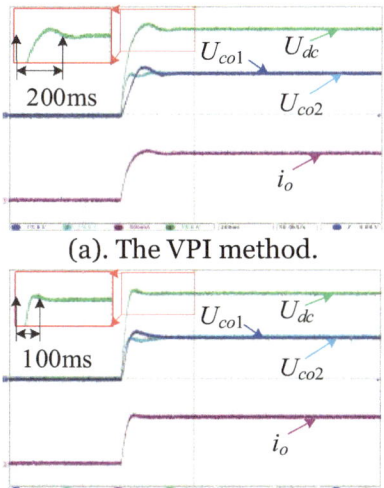

(a). The VPI method.

(b). The tunable power sharing control strategy.

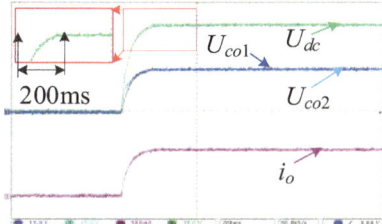

(c). The variant tunable power sharing control scheme.

Fig. 4.15 Experimental results of start-up process under different methods. (U_{co1}, U_{co2} and U_{dc}: 15 V/div; i_o: 0.5 A/div; Time: 200 ms/div)

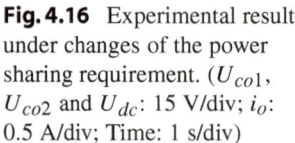

Fig. 4.16 Experimental results under changes of the power sharing requirement. (U_{co1}, U_{co2} and U_{dc}: 15 V/div; i_o: 0.5 A/div; Time: 1 s/div)

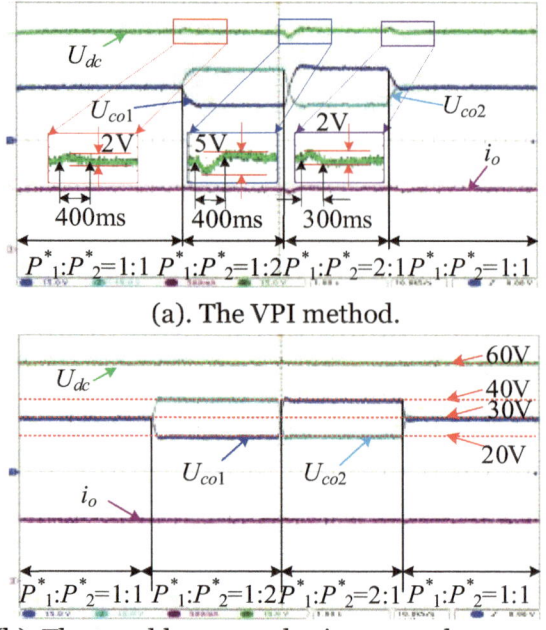

obtain their desired values simultaneously, and the black-start performance of the IPOS DAB dc-dc converter can be implemented.

When R is set to 56 Ω, Fig. 4.16 shows the experimental results under disturbances of power sharing coefficients. When the power sharing requirement is set as $P^*_1:P^*_2 = 1:1$, U_{co1} and U_{co2} in steady state condition should be stabilized at 30 V and 30 V, respectively, and when output power of one DAB module is required as double as the output power of the other DAB module, its output voltage should also be twice the other's output voltage. As shown in Fig. 4.16a, the disturbances of U_{dc} are bigger than 2 V during the adjusting process of U_{co1} and U_{co2} under VPI method, and the settling time of U_{dc} is about 400 ms for a new power sharing performance. Moreover, based on the tunable power sharing control strategy, U_{dc} can stay stable when adjusting U_{co1} and U_{co2} (see Fig. 4.16b).

When R is set to 56 Ω, Fig. 4.17 shows the experimental results during plug-out and plug-in processes of the second DAB module (representing the second ESS equipment). As shown in Fig. 4.17a, when the second DAB module is plugged-out or plugged-in, the disturbance of U_{dc} is bigger than 4 V under VPI method, and the settling time for both processes is greater than 100 ms. Moreover, as shown in Fig. 4.17b, regardless of plug-out process or plug-in process, U_{dc} can be maintained at its desired using the presented tunable power sharing control strategy.

Fig. 4.17 Experimental results during the hot swap process. (U_{co1}, U_{co2} and U_{dc}: 15 V/div; i_o: 0.5 A/div; Time: 1 s/div)

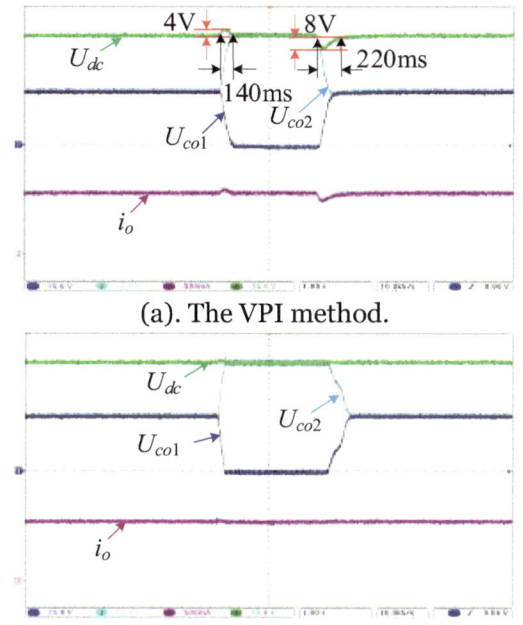

(a). The VPI method.

(b). The tunable power sharing control strategy.

Fig. 4.18 Experimental results under disturbances of load resistor. (U_{co1}, U_{co2} and U_{dc}: 15 V/div; i_o: 0.5 A/div; Time: 1 s/div)

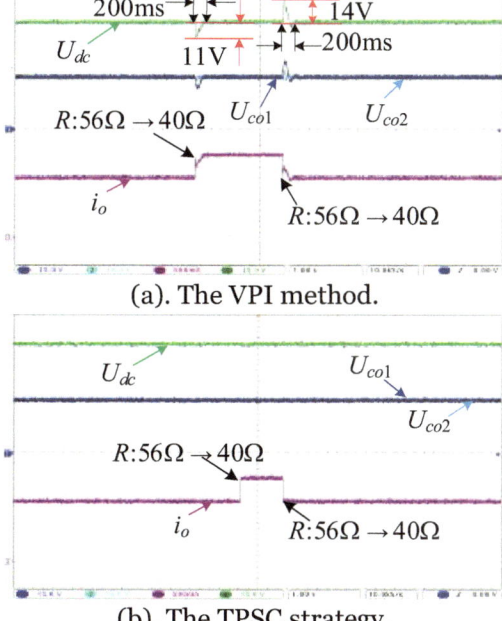

(a). The VPI method.

(b). The TPSC strategy.

Figure 4.18 shows experimental results of the IPOS DAB dc-dc converter under various load resistors between 40 and 56 Ω. As shown in Fig. 4.18a, disturbances of U_{dc}, U_{co1} and U_{co2} are present under VPI method, and change in U_{dc} is close to 12 V. Moreover, the settling time of U_{dc} is about 200 ms. In addition, based on the tunable power sharing control strategy, U_{dc}, U_{co1} and U_{co2} are maintained at their desired values under tunable power sharing control scheme (see from Fig. 4.18b), and the high robustness for the IPOS DAB dc-dc converter can be provided.

4.3.2 A Power Sharing Control Scheme with Fast-Dynamic Response for the ISOP DAB DC-DC Converter System

In this section, based on the circuit parameters in Table 4.2, a simulation model and a small-scale experiment platform with two DAB modules is built to verify the proposed power sharing control method with fast-dynamic response for the ISOP DAB dc-dc converter system.

A. *The Simulation Results.*

The comparison of the traditional method [16, 17] and the proposed power sharing control scheme with fast-dynamic response is provided. Moreover, by using the proposed general inductance-estimating method for the ISOP DAB dc-dc converter, the inductance values under the traditional method are estimated with the measurement of the load current.

When the input voltage is 100 V and the power balance performance is achieved, Fig. 4.19 shows the simulation results of the traditional method and the proposed scheme when the load resistor is changed between 15 and 50 Ω. As shown in Fig. 4.19b, when the load resistor is changed, the total transferred current $\sum i_{TX}$ under the proposed scheme can follow with the change of the load current timely as shown in Fig. 4.19a. Thus, as shown in Fig. 4.19c, the output voltage can remain at its desired value 50 V by using the proposed scheme. However, compared with Fig. 4.19a and b under the traditional method, the total transferred current cannot be stable at the required current timely, so the output voltage disturbances are obvious.

Moreover, when the input voltage is 100 V and the load resistor is 15 Ω, Fig. 4.20 shows the simulation results when the power sharing performance of these two DAB modules is changed between 1:1 and 2:1 (see from Fig. 4.20a and b). As shown in Fig. 4.20c, during the regulation of input voltages, the transferred current under the proposed scheme can keep constant, but the transferred current under the traditional method has some disturbances. Besides, according to Fig. 4.20d, the stable output voltage can be obtained by using the proposed scheme, but the output-voltage disturbances under the traditional method is obvious.

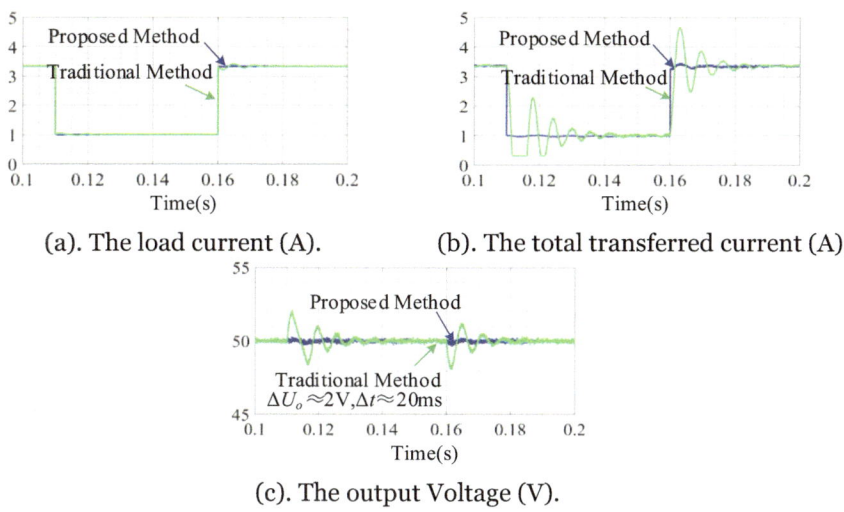

(a). The load current (A).   (b). The total transferred current (A).

(c). The output Voltage (V).

Fig. 4.19 The simulation results when the load resistor is changed. (t: 20 ms/div)

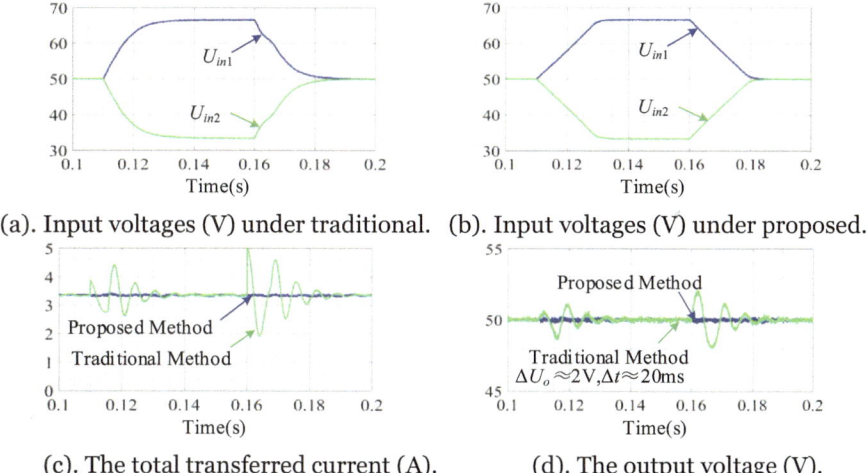

(a). Input voltages (V) under traditional.   (b). Input voltages (V) under proposed.

(c). The total transferred current (A).   (d). The output voltage (V).

Fig. 4.20 The simulation results when the power sharing performance is changed. (t: 20 ms/div)

In addition, when L_1 is changed to 30μH in the simulation model, Fig. 4.21 shows the simulation results of the proposed inductance-estimating method embedded in the traditional method [18, 19] at the cost of an additional load-current sensor. As shown in Fig. 4.21a, the estimated inductances are close to the configured values. The relationship

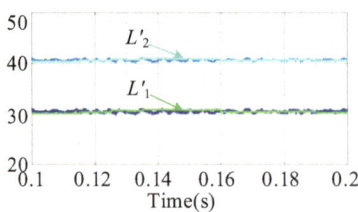

(a). Estimated inductances (μH).

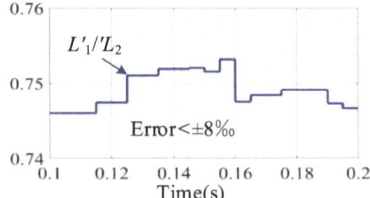

(b). The relationship of L'_1 and L'_2.

Fig. 4.21 The simulation results of the estimated inductance by using the proposed general inductance-estimating method embedded in the traditional control method. (t: 20 ms/div)

of these two estimated inductances can be accurate as shown in Fig. 4.21b. Therefore, the proposed general inductance-estimating method can also be employed in other control schemes for the ISOP DAB dc-dc converter if the accuracy of the inductances should be ensured.

B. *The Experiment Results*

When the inductance of the first DAB module L_1 is inaccurate as 24μH, Fig. 4.22 shows the experiment results when the input voltage and the load resistor are changed. As shown in Fig. 4.22a and b, when the total input voltage is changed between 100 and 120 V, the output voltage U_o can be kept at its desired value as 50 V. However, the power balance performance cannot be achieved since the inductance value L_1 is not accurate. Moreover, as shown in Fig. 4.22c and d, when the load resistor is changed between 15 and 50 Ω, the output voltage U_o can be stable at its desired value. Therefore, when the inductance value is not accurate, the desired power sharing performance cannot be realized, but the excellent dynamic response can be provided for the ISOP DAB dc-dc converter.

When the total input voltage is 100 V, the desired output voltage is 50 V and the load resistor is 15 Ω, the experiment result of these estimated inductances of these two DAB modules can be shown in Fig. 4.23. As shown in Fig. 4.23, the estimated inductance of the first DAB module L_1 is 43.2 μH, and the estimated inductance of the second DAB module L_2 is 43.7 μH. Thus, these estimated inductances are close to the actual inductances as shown in Table 4.2 but a little bigger, which should be caused by the power losses.

Based on the estimated inductances, the experiment results can be shown in Fig. 4.24 when the input voltage, the load resistor, and the power sharing performance are changed. A shown in Fig. 4.24a and b, the power balance performance of the ISOP DAB dc-dc converter can be ensured, and when the input voltage is changed, the output voltage can be stable at its desired value. Moreover, as shown in Fig. 4.24c and b, when the load resistor is changed between 15 and 50 Ω, the output-voltage disturbance can be regarded as zero. In addition, As shown in Fig. 4.24e and c, when the desired power sharing performance

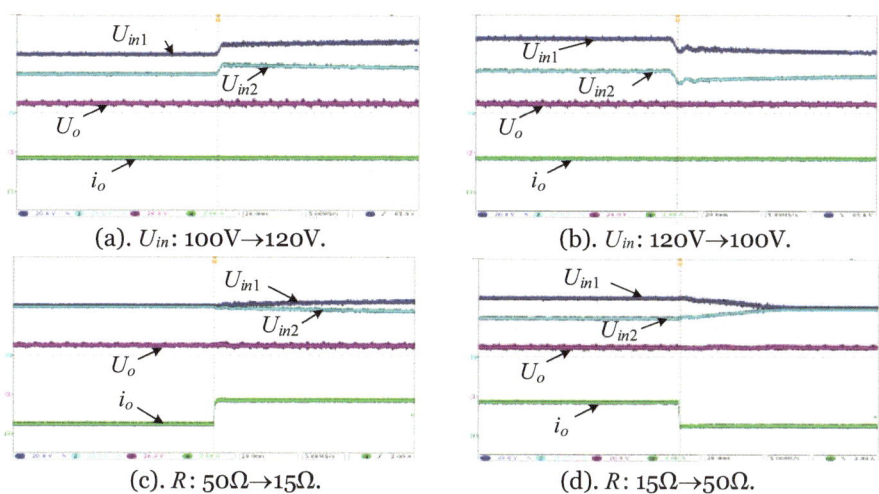

(a). U_{in}: 100V→120V.   (b). U_{in}: 120V→100V.

(c). R: 50Ω→15Ω.   (d). R: 15Ω→50Ω.

Fig. 4.22 The experiment results when $L_1 = 24\,\mu H$ and $L_2 = 40\mu H$. (U_{in1}, U_{in2} and U_o: 20 V/div; i_o: 2 A/div; t: 20 ms/div)

Fig. 4.23 The estimated inductances

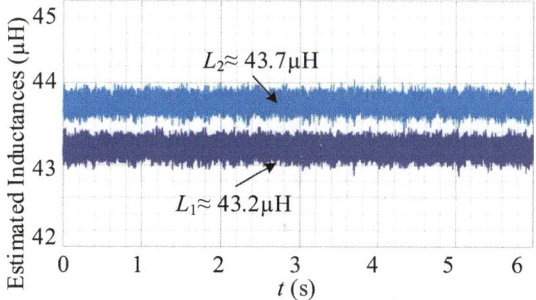

of these two DAB modules is changed between 1:1 and 2:1, the corresponding actual power sharing performances are 1:1 and 1.97:1. So, based on the proposed scheme, the power sharing performance of the ISOP DAB dc-dc converter can be adjusted flexibly. Furthermore, when the power sharing performance is changed, the output voltage can be kept at its desired value. Therefore, with the estimated inductances, the required power sharing performance can be ensured, and the excellent dynamic response can be achieved when the input voltage, the load resistor, and the power sharing performance are changed.

With the actual inductances of these two DAB modules, the experiment results, when the input voltage, the load resistor and the power sharing performance are changed, can be shown in Fig. 4.25. Similarly, when the input voltage, the load resistor and the power sharing performance of the ISOP DAB dc-dc converter are changed, the output voltage can be kept at its desired value, and the excellent dynamic performance can be provided

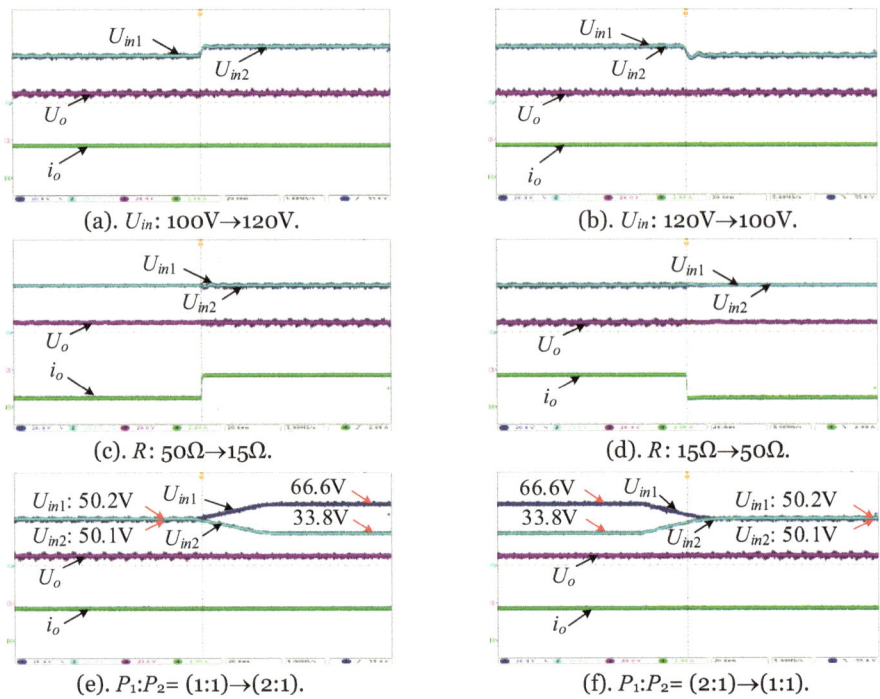

(a). U_{in}: 100V→120V.

(b). U_{in}: 120V→100V.

(c). R: 50Ω→15Ω.

(d). R: 15Ω→50Ω.

(e). $P_1:P_2$= (1:1)→(2:1).

(f). $P_1:P_2$= (2:1)→(1:1).

Fig. 4.24 The experiment results with estimated inductances $L_1 = 43.2\mu\mathrm{H}$ and $L_2 = 43.7\mu\mathrm{H}$. (U_{in1}, U_{in2} and U_o: 20 V/div; i_o: 2 A/div; t: 20 ms/div)

for this converter system. Moreover, as shown in Fig. 4.25e and f, when the desired output power of the first DAB module is double as that of the second DAB module, the actual power sharing performance of these two modules is 1.99: 1, and when the desired output power of the first DAB modules should be the same as that of the second DAB module, this power balance requirement can be achieved. Therefore, with actual inductance values, the desired power sharing performance can be ensured for the ISOP DAB dc-dc converter by using the proposed power sharing control scheme with fast-dynamic response. Thus, compared Figs. 4.24 with 4.25, the effectiveness of the proposed inductance-estimating method can be verified.

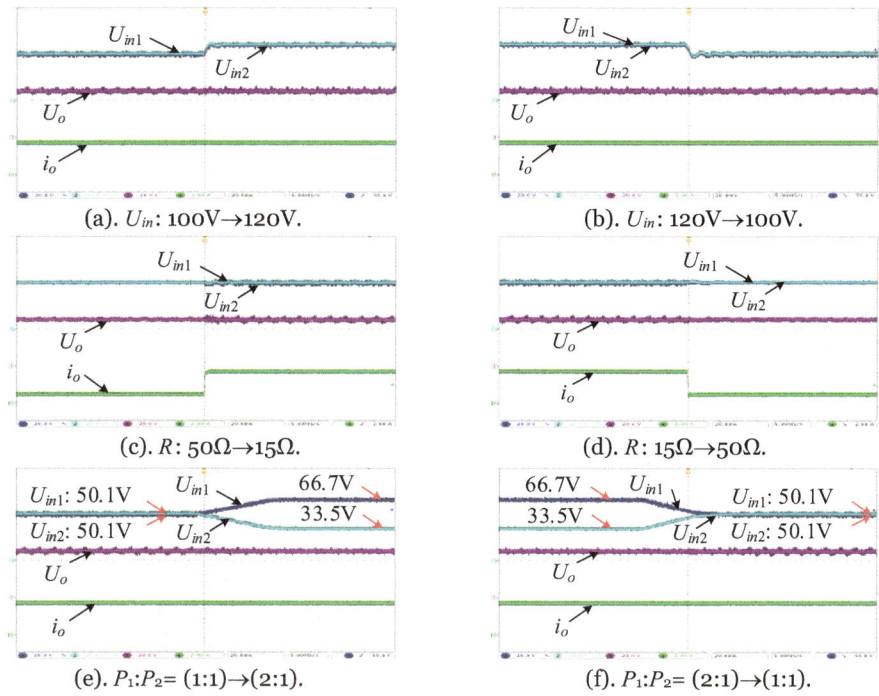

(a). U_{in}: 100V→120V.

(b). U_{in}: 120V→100V.

(c). R: 50Ω→15Ω.

(d). R: 15Ω→50Ω.

(e). P_1:P_2= (1:1)→(2:1).

(f). P_1:P_2= (2:1)→(1:1).

Fig. 4.25 The experiment results when $L_1 = 40$ µH and $L_2 = 40$ µH. (U_{in1}, U_{in2} and U_o: 20 V/div; i_o: 2 A/div; t: 20 ms/div)

4.4 Summary

In this chapter, a tunable power sharing control strategy is proposed for the IPOS DAB dc-dc converter system. Based on this scheme, the fast-dynamic control performance can be obtained when the input voltage, the load resistor and the power sharing performance are changed. Besides, with a simple variation, the tunable power sharing control strategy can implement black-start performance for the IPOS DAB dc-dc converter system. Moreover, based on the presented hot-swap operations, when one DAB module is plugged-in or -out, the output-side total voltage can stay unchanged. In addition, an input-oriented power sharing control scheme with fast-dynamic response is proposed for the ISOP DAB dc-dc converter system, which can realize the complete decoupling between the regulation of the input voltage and the adjustment of output voltage. Besides, the proposed scheme can provide an excellent dynamic response for the ISOP DAB dc-dc converter system when the total input voltage, the load resistor and the power sharing performance are changed. Furthermore, based on the proposed inductance-estimating method, the inductance values for each DAB module can be estimated, which can also be employed to ensure the power sharing performance of the ISOP DAB dc-dc converter system.

References

1. N. Hou, W. Song, Y. Li, Y. Zhu, and Y. Zhu, "A Comprehensive Optimization Control of Dual-Active-Bridge DC–DC Converters Based on Unified-Phase-Shift and Power-Balancing Scheme," *IEEE Transactions on Power Electronics,* vol. 34, no. 1, pp. 826–839, 2019.
2. R. W. A. A. D. Doncker, D. M. Divan, and M. H. Kheraluwala, "A three-phase soft-switched high-power-density DC/DC converter for high-power applications," *IEEE Transactions on Industry Applications,* vol. 27, no. 1, pp. 63–73, 1991.
3. Y. Zhang, and Y. W. Li, "Energy Management Strategy for Supercapacitor in Droop-Controlled DC Microgrid Using Virtual Impedance," *IEEE Transactions on Power Electronics,* vol. 32, no. 4, pp. 2704–2716, 2017.
4. L. Maharjan, S. Inoue, H. Akagi, and J. Asakura, "State-of-Charge (SOC)-Balancing Control of a Battery Energy Storage System Based on a Cascade PWM Converter," *IEEE Transactions on Power Electronics,* vol. 24, no. 6, pp. 1628–1636, 2009.
5. K. Bi, L. Sun, Q. An, and J. Duan, "Active SOC Balancing Control Strategy for Modular Multilevel Super Capacitor Energy Storage System," *IEEE Transactions on Power Electronics,* vol. 34, no. 5, pp. 4981–4992, 2019.
6. L. Hui, P. Fang Zheng, and J. S. Lawler, "A natural ZVS medium-power bidirectional DC-DC converter with minimum number of devices," *IEEE Transactions on Industry Applications,* vol. 39, no. 2, pp. 525–535, 2003.
7. A. J. B. Bottion, and I. Barbi, "Input-Series and Output-Series Connected Modular Output Capacitor Full-Bridge PWM DC–DC Converter," *IEEE Transactions on Industrial Electronics,* vol. 62, no. 10, pp. 6213–6221, 2015.
8. S. P. Engel, N. Soltau, H. Stagge, and R. W. D. Doncker, "Dynamic and Balanced Control of Three-Phase High-Power Dual-Active Bridge DC–DC Converters in DC-Grid Applications," *IEEE Transactions on Power Electronics,* vol. 28, no. 4, pp. 1880–1889, 2013.
9. W. Song, N. Hou, and M. Wu, "Virtual Direct Power Control Scheme of Dual Active Bridge DC–DC Converters for Fast Dynamic Response," *IEEE Transactions on Power Electronics,* vol. 33, no. 2, pp. 1750–1759, 2018.
10. N. Hou, Y. Li, Z. Quan, Y. W. Li, and A. Zhou, "Unified Fast-Dynamic Direct-Current Control Scheme for Intermediary Inductive AC-Link Isolated DC-DC Converters," *IEEE Open Journal of Power Electronics,* vol. 2, pp. 383–400, 2021.
11. V. M. Iyer, S. Gulur, and S. Bhattacharya, "Small-Signal Stability Assessment and Active Stabilization of a Bidirectional Battery Charger," *IEEE Transactions on Industry Applications,* vol. 55, no. 1, pp. 563–574, 2019.
12. N. Hou, W. Song, and M. Wu, "Minimum-Current-Stress Scheme of Dual Active Bridge DC–DC Converter With Unified Phase-Shift Control," *IEEE Transactions on Power Electronics,* vol. 31, no. 12, pp. 8552–8561, 2016.
13. J. W. Kimball, J. T. Mossoba, and P. T. Krein, "A Stabilizing, High-Performance Controller for Input Series-Output Parallel Converters," *IEEE Transactions on Power Electronics,* vol. 23, no. 3, pp. 1416–1427, 2008.
14. M. Abrehdari, and M. Sarvi, "Comprehensive sharing control strategy for input-series output-parallel connected modular DC–DC converters," *IET Power Electronics,* vol. 12, no. 12, pp. 3105–3117, 2019.
15. L. Qu, D. Zhang, and Z. Bao, "Output Current-Differential Control Scheme for Input-Series–Output-Parallel-Connected Modular DC–DC Converters," *IEEE Transactions on Power Electronics,* vol. 32, no. 7, pp. 5699–5711, 2017.

16. P. Zumel, L. Ortega, A. Lázaro, C. Fernández, A. Barrado, A. Rodríguez, and M. M. Hernando, "Modular Dual-Active Bridge Converter Architecture," *IEEE Transactions on Industry Applications,* vol. 52, no. 3, pp. 2444–2455, 2016.

17. C. Luo, and S. Huang, "Novel Voltage Balancing Control Strategy for Dual-Active-Bridge Input-Series-Output-Parallel DC-DC Converters," *IEEE Access,* vol. 8, pp. 103114–103123, 2020.

18. J. Liu, J. Yang, J. Zhang, Z. Nan, and Q. Zheng, "Voltage Balance Control Based on Dual Active Bridge DC/DC Converters in a Power Electronic Traction Transformer," *IEEE Transactions on Power Electronics,* vol. 33, no. 2, pp. 1696–1714, 2018.

19. L. Wang, Q. Zhu, W. Yu, and A. Q. Huang, "A Medium-Voltage Medium-Frequency Isolated DC–DC Converter Based on 15-kV SiC MOSFETs," *IEEE Journal of Emerging and Selected Topics in Power Electronics,* vol. 5, no. 1, pp. 100–109, 2017.

The DAB-Based PPP Converter System with Robust DC-Link Voltage for Islanded DC Microgrid

5

In this chapter, based on SPS modulation, the topology of this DAB-based PPP converter system is analyzed in Sect. 5.1. Moreover, to boost the robustness of the dc-link voltage, a high-robustness control strategy is proposed for maintaining the dc-link voltage when the working condition of the renewable energy source, the output voltage of the battery and the load condition are changed in Sect. 5.2. Besides, when one renewable source is out of work, the corresponding operation is presented. Finally, simulation results and experiment results are provided to verify the effectiveness of the proposed DAB-based PPP converter system and the proposed high-robustness control scheme with fast-dynamic response in Sect. 5.3. Then, the logic structure of this chapter can be summarized in Fig. 5.1.

5.1 The DAB-Based PPP Converter System

In this section, the simplified circuit of the DAB-based PPP converter system as shown in Fig. 1.15 is analyzed at first, and the SPS modulation method is discussed for realizing the bidirectional power transmission of the converter system.

5.1.1 The Simplified Circuit of the DAB-Based PPP Converter System

The DAB-based PPP converter system can be shown in Fig. 1.15. Since the transferred current of the DAB dc-dc converter can be determined by the circuit parameter and the phase-shift ratio, the middle ac inductance can be neglectful [1, 2]. So, the DAB converter

© The Author(s), under exclusive license to Springer Nature Switzerland AG 2025
N. Hou, *High-Robust Control Schemes for Dual-Active-Bridge-Based DC–DC Converter Systems in Renewable Energy Applications*, Synthesis Lectures on Power Electronics,
https://doi.org/10.1007/978-3-031-72963-8_5

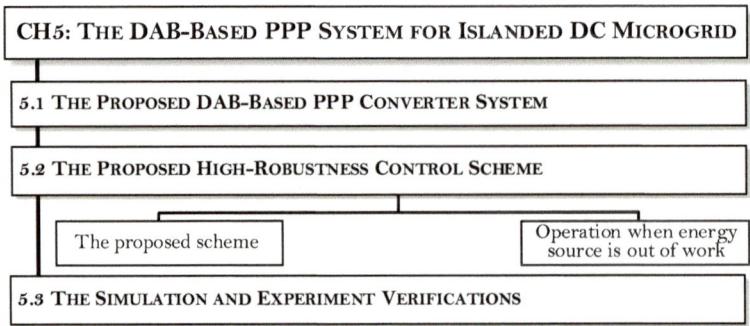

CH5: The DAB-Based PPP System for Islanded DC Microgrid

5.1 The Proposed DAB-Based PPP Converter System

5.2 The Proposed High-Robustness Control Scheme

The proposed scheme

Operation when energy source is out of work

5.3 The Simulation and Experiment Verifications

Fig. 5.1 Logic structure of this chapter

can be treated as a controllable current source by using different control values. Moreover, renewable energy source such as the PV panel, fuel cell and WT with ac-dc conversion can be modeled as a current source [3, 4], and the battery is a voltage source. Then, by using PV as an example, the simplified circuit of the DAB-based PPP converter system can be shown in Fig. 5.2. According to Fig. 5.2, the average model of the DAB-based converter system can be expressed as,

$$
\begin{cases}
C_{in\alpha} \frac{dU_{PV\alpha}}{dt} = i_{PV\alpha} - i_{bus} - \frac{I_{T\alpha}U_{LD}}{U_{PV\alpha}} \\
C_{inm} \frac{dU_C}{dt} = -\frac{I_{Tm}U_{LD}}{U_C} - i_{bus} \\
C_{MD} \frac{dU_{MD}}{dt} = i_{bus} - i_{MD} \\
U_{MD} = U_C + \sum_{\alpha=1}^{m-1} U_{PV\alpha}
\end{cases}
\quad (\alpha \in [1, m-1]) \qquad (5.1)
$$

Fig. 5.2 The simplified circuit of the DAB-based PPP converter system with battery integration

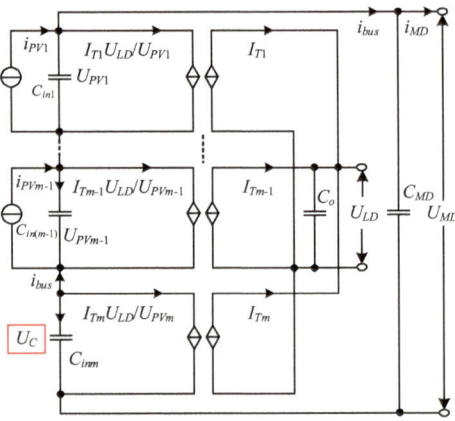

Based on (4.1), the transferred current of the first $(m\text{-}1)^{\text{th}}$ DAB modules can be employed to adjust the output voltage $U_{PV\alpha}$ of the PV panels for the MPPT performance. Besides, to maintain the dc-link voltage U_{MD}, the input voltage U_C of the compensated DAB module should be controllable, which can be realized by adjusting the transferred current I_{TM} of this DAB module.

5.1.2 The Single-Phase-Shift Modulation Method

For the DAB dc-dc converter, the SPS modulation method is the most popular modulation method for realizing flexible power transmission. Thus, the SPS modulation method is adopted, which can be illustrated in Fig. 5.3 for bidirectional power transmission, where $U_{ab\alpha}$ is the output voltage of the primary-side H Bridge, $U_{cd\alpha}$ is the output voltage of the secondary-side H Bridge, $i_{L\alpha}$ is the inductance current, D_α is the phase-shift ratio and $T_{s\alpha}$ is the switching period of the corresponding DAB modules.

According to Fig. 5.3, the transferred power P_α of DAB module connected to PV panel under SPS modulation method can be expressed as,

$$P_\alpha = \begin{cases} \dfrac{U_{PV\alpha}U_{LD}D_\alpha(1-D_\alpha)T_{s\alpha}}{2n_\alpha L_\alpha} & (P_\alpha \geq 0) \\[3mm] -\dfrac{U_{PV\alpha}U_{LD}D_\alpha(1-D_\alpha)T_{s\alpha}}{2n_\alpha L_\alpha} & (P_\alpha < 0) \end{cases} \tag{5.2}$$

The transferred current $I_{T\alpha}$ of the DAB module can be expressed as,

$$I_{T\alpha} = \dfrac{P_\alpha}{U_{LD}} = \begin{cases} \dfrac{U_{PV\alpha}D_\alpha(1-D_\alpha)T_{s\alpha}}{2n_\alpha L_\alpha} & (I_{T\alpha} \geq 0) \\[3mm] -\dfrac{U_{PV\alpha}D_\alpha(1-D_\alpha)T_{s\alpha}}{2n_\alpha L_\alpha} & (I_{T\alpha} < 0) \end{cases} \tag{5.3}$$

According to (5.3), the phase-shift ratio D_α can be calculated as,

Fig. 5.3 The SPS modulation method of the DAB converter for bidirectional power flowing conditions

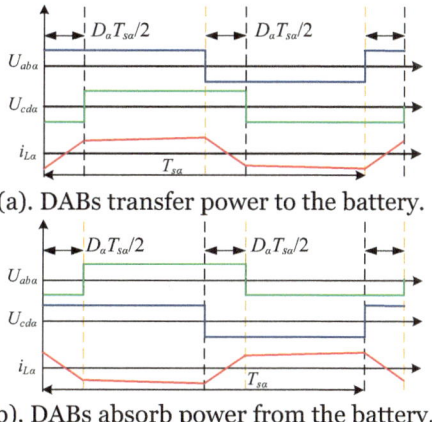

(a). DABs transfer power to the battery.

(b). DABs absorb power from the battery.

$$D_\alpha = \begin{cases} \frac{1}{2} - \sqrt{1 - \frac{8L_\alpha I_{T\alpha}}{n_\alpha U_{PV\alpha} T_{s\alpha}}} & (I_{T\alpha} \geq 0) \\ \frac{1}{2} - \sqrt{1 + \frac{8n_\alpha L_\alpha I_{T\alpha}}{U_{PV\alpha} T_{s\alpha}}} & (I_{T\alpha} < 0) \end{cases} \tag{5.4}$$

Similarly, the phase-shift ratio D_m of the mth DAB module can be expressed by its transferred current I_{Tm} as,

$$D_m = \begin{cases} \frac{1}{2} - \sqrt{1 - \frac{8L_m I_{Tm}}{n_m U_C T_{sm}}} & (I_{Tm} \geq 0) \\ \frac{1}{2} - \sqrt{1 + \frac{8n_m L_m I_{Tm}}{U_C T_{sm}}} & (I_{Tm} < 0) \end{cases} \tag{5.5}$$

5.2 The Proposed High-Robustness Control Strategy

In this section, a high-robustness control strategy is proposed for the DAB-based PPP converter system with battery integration as Fig. 1.15, which can provide the required output voltage for the renewable energy source and boost the robustness of the total dc-link voltage. By using the PV as an example, the MPPT performance can be realized with the required output voltage. Moreover, when the renewable energy source is out of work, the corresponding operation is also presented based on the high-robustness control scheme, which can be employed when the PV module is heavily shaded or at night for all the DAB converters connected with PV panels.

5.2.1 The High-Robustness Control Scheme

For the PV panels, the typical output characteristic can be shown in Fig. 5.4, and under different irradiance, there is a maximum output power point [4, 5]. By adjusting the output voltage of the PV panel, this MPPT performance can be obtained, and these output voltages under different irradiances are relatively close, which also benefits the proposed PPP converter system. Moreover, perturb and observe (P&O) method or the hill climbing method is the widely used technique to track the maximum output power of PV panels as shown in Fig. 5.5 [6, 7]. By positively perturbing the output voltage of the PV panel, the difference in power before and after perturbation can be obtained. If the power difference is positive, the direction of the perturbation should be the same. Otherwise, the perturbating direction should be reversed.

Moreover, according to Fig. 5.2, to meet the requirement of the load side, the bus current i_{bus} should be equivalent to the load current i_{MD}. To immediately track the change of load, the required current i^*_{MD} of each DAB module for supporting the electricity consumption can be expressed as,

$$i^*_{MD} = \frac{U^*_{MD} i_{MD}}{U_{MD}} \tag{5.6}$$

Fig. 5.4 The typical output characteristic of PV panel

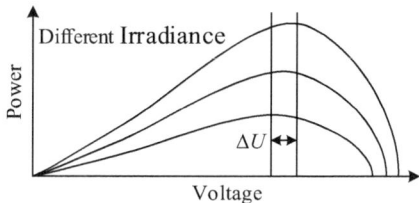

Fig. 5.5 The basic block diagram of P&O MPPT

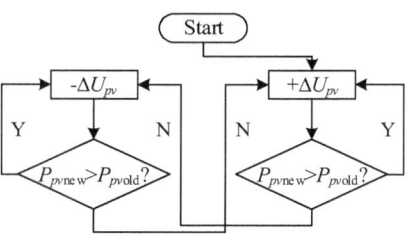

Combining (4.1), (3.38)~(5.6) and Fig. 5.5, the proposed high-robustness control scheme for the DAB-based PPP converter system with battery integration can be illustrated in Fig. 5.6. At the beginning of the switching period, the output current of each PV panel, the voltage of each PV panel, the input voltage of the additional converter, the load current, the battery voltage, and the total dc-link voltage are measured. Moreover, based on the output voltage and the output current of each PV panel in the current switching period and the last switching period, the new desired voltage of each PV panel can be obtained by using the P&O MPPT scheme [6, 7], and the other MPPT schemes can also be employed to obtained the desired voltage of each PV panel. Similarly, the required output voltage for fuel cell can be obtained from the control of the output current, and the constant voltage is enough for the wind turbine with ac-dc conversion.

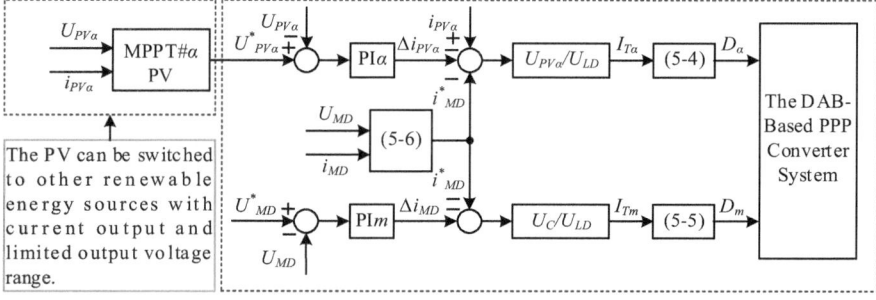

Fig. 5.6 The control system for the partial power processing converter system with robustness dc-link voltage

In addition, based on the load current and the dc-link voltage, the total desired output current for supporting the load side can be obtained. For the DAB module connected with the PV panel, the total transferred current can be obtained by combining the current which is employed to adjust the voltage of PV panel. This current can be obtained through the corresponding PI controller with the PV voltage and the desired voltage. Similarly, by combining the current for maintaining the dc-link voltage from the mth PI controller, the transferred current for the additional DAB dc-dc converter can be obtained. Besides, based on (5.4) and (5.5), the required phase-shift ratio for each DAB module can be obtained, which can be used to realize the control of the DAB-based PPP converter system. Thus, the proposed high robustness control scheme can be realized for controlling this DAB-based PPP converter system with the independent control of the renewable energy source and the stabilization of the total dc-link voltage. Furthermore, the PV panels can be switched to other renewable energy sources with current output and limited output voltage range, the desired terminal voltage can be obtained from the outer control loop for these renewable energy sources.

5.2.2 The Operation When One Renewable Energy Source is Out of Work

Sometimes, when the renewable energy source is out of work, the corresponding operation should be provided to determine the desired terminal voltage of the related DAB module. For example, when one PV module is heavily shaded, the output current of this PV panel is very small. Moreover, the output current of the PV panel should be zero at night. Then, the MPPT becomes meaningless. Thus, to deal with these conditions, these DAB modules can be turned into constant voltage control, and the desired voltage will be constant after the output current of the PV panel is smaller than the lower limit. Then, the simplified circuit of the DAB-based converter system can be shown in Fig. 5.7.

5.3 Verification

In this section, by using the PV panels as example, a simulation model with three DAB modules and a small-scale experiment platform with two DAB modules is built to verify the effectiveness of the proposed DAB-based PPP system with battery integration and the proposed high-robustness control strategy with PV-covered operation for this converter system.

Fig. 5.7 The simplified circuit of the DAB-based converter system when one renewable energy source such as PV is out of work

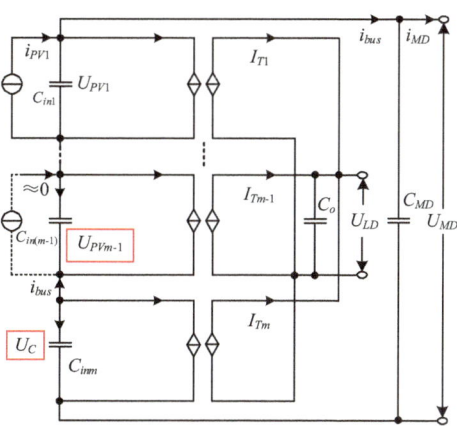

5.3.1 The Simulation Results with Two PV Panels and One Compensating Module

A simulation model with three DAB modules is built to verify the effectiveness of the proposed DAB-based PPP converter system and the proposed high robustness control scheme with fast-dynamic response for this converter system. Since the battery can be also switched into a low-voltage dc bus, a dc voltage source is employed to replace the integrated battery for testing the function of this proposed converter and the corresponding control method. The circuit parameters of the converter system with three modules are shown in Table 5.1.

When the battery voltage U_{LD} is 90 V and the load resistor R is 150 Ω, Fig. 5.8 shows the simulation result with changed irradiances of PV panels. As shown in Fig. 5.8a, the irradiance of the first PV panel is changed from 600 to 560 W/m^2 then to 600 W/m^2, and the irradiance of the second PV panel is increased from 520 to 560 W/m^2 then to 600

Table 5.1 Circuit parameters of the simulation model

Parameter	Value
L_1, L_2, L_3	4 mH
n_1, n_2, n_3	1/5
f_s	1 kHz
R	150~300 Ω
U_{LD}	80~90 V
U^*_{MD}	1000 V
Modules of PV1 in series	14
Modules of PV2 in series	13
PV1, PV2	Trina solar TSM-250PA05.08

W/m^2. Moreover, the desired output voltages of PV panels can be shown in Fig. 5.8b, and the corresponding output current of PV panels can be shown in Fig. 5.8c. Based on the proposed high robustness control scheme, the output voltages of PV panels and the compensated voltage can be shown in Fig. 5.8d, and the total dc-link voltage U_{MD} can be obtained as shown in Fig. 5.8e. Thus, based on the proposed control method, the robustness of the total dc-link voltage can be ensured when the irradiances of PV panels are changed.

When the irradiance of the first PV panel is 600 W/m^2 and the irradiance of the second PV panel is 520 W/m^2, Fig. 5.9 shows the simulation result with a changed load resistor. As shown in Fig. 5.9a, the load resistor is changed between 150 and 300 Ω. Based on the proposed high robustness control scheme, the output voltages of PV panels and the compensated voltage can be shown in Fig. 5.9b, and the total dc-link voltage U_{MD} can be obtained as shown in Fig. 5.9c, where the total dc-link voltage is maintained at its desired value. Therefore, based on the proposed control method, excellent dynamic performance can be provided for the presented DAB-based PPP converter system when the load resistor is changed.

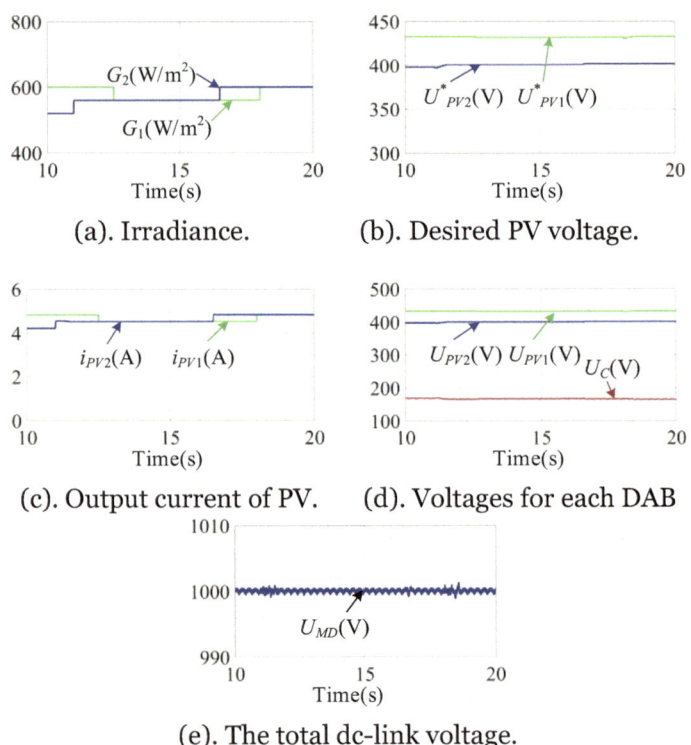

(a). Irradiance.   (b). Desired PV voltage.

(c). Output current of PV.   (d). Voltages for each DAB

(e). The total dc-link voltage.

Fig. 5.8 Simulation results when the irradiances are changed

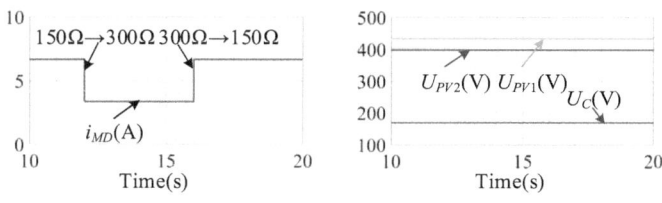

(a). Output current of dc bus. (b). Voltages for each DAB.

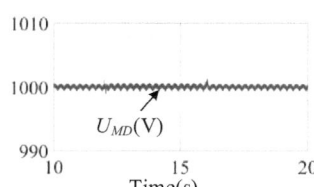

(c). The total dc-link voltage.

Fig. 5.9 Simulation results when load resistor is changed

When the irradiance of the first PV panel is 600 W/m^2 and the irradiance of the second PV panel is 520 W/m^2, Fig. 5.10 shows the simulation result with a changed voltage of the low-voltage bus. As shown in Fig. 5.10a, the voltage of the low-voltage bus is changed between 80 and 90 V. Based on the proposed high robustness control scheme, the output voltages of PV panels and the compensated voltage can be shown in Fig. 5.10b. Besides, as shown in Fig. 5.10c, the total dc-link voltage can be kept at its desired value by using the proposed high robustness control scheme. Therefore, based on the proposed control method, excellent dynamic performance can be provided for the presented DAB-based PPP converter system when the voltage of the low voltage terminal is changed.

When the battery voltage U_{LD} is 90 V and the load resistor R is 150 Ω, Fig. 5.11 shows the simulation result when the PV panels are heavily covered sometimes. As shown in Fig. 5.11a, the irradiance of the first PV panel is become as zero at first and then returned to 600 W/m^2 then to 600 W/m^2, and the irradiance of the second PV panel is reduced to zero without recovery. Moreover, combining the operation for the heavily shaded PV panel, the desired output voltages of PV panels can be shown in Fig. 5.11b, and the corresponding output current of PV panels can be shown in Fig. 5.11c. Based on the proposed high robustness control scheme, the output voltages of PV panels and the compensated voltage can be shown in Fig. 5.11d, and the total dc-link voltage U_{MD} can be obtained as shown in Fig. 5.11e. Thus, based on the proposed control method, the robustness of the total dc-link voltage can be ensured even when the PV panels are heavily shaded.

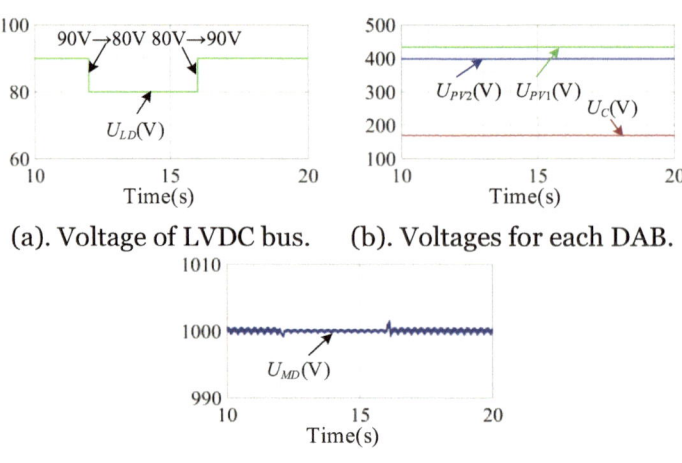

(a). Voltage of LVDC bus.   (b). Voltages for each DAB.

(c). The total dc-link voltage.

Fig. 5.10 Simulation results when the voltage of LVDC is changed

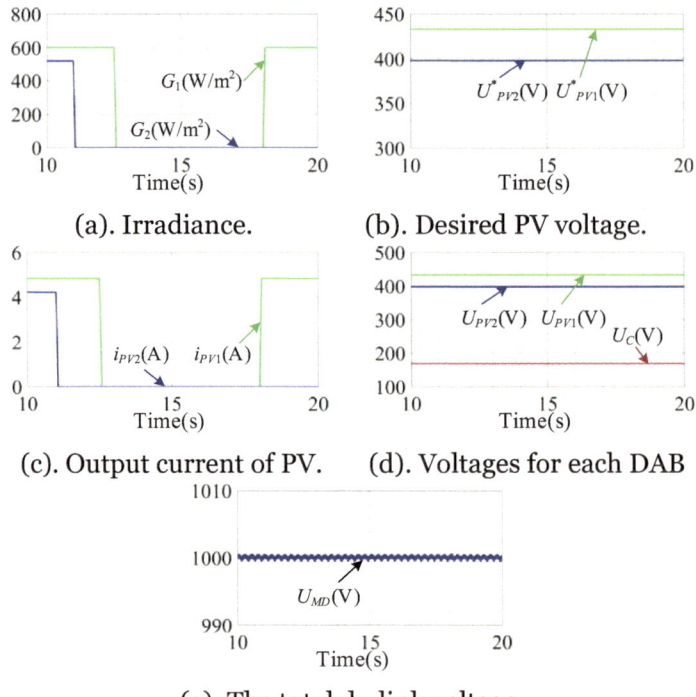

(a). Irradiance.   (b). Desired PV voltage.

(c). Output current of PV.   (d). Voltages for each DAB

(e). The total dc-link voltage.

Fig. 5.11 Simulation results when the PV panel is covered

5.3.2 The Experiment Results with One PV Panel and One Compensating Module

A small-scale experiment platform with two DAB modules is employed to verify the effectiveness of the proposed DAB-based PPP converter system and the proposed high robustness with fast-dynamic response for this converter system. The main circuit parameters of this experiment platform can be shown in Table 5.2. Moreover, the configuration of the small-scale experiment platform can be shown in Fig. 5.12, where the power supply Agilent E4360A with the changeable output current and the adjustable output voltage is employed to simulate the PV panel and the power supply Sorensen SGX60X83C is used to replace the battery. As shown in Fig. 5.12, the total dc-link voltage U_{MD}, the load current i_{MD}, the output current i_{PV} of PV, the terminal voltage U_{PV} and the voltage U_{LD} are measured by an oscilloscope.

When the load resistor is 40 Ω, the voltage of the LVDC bus is 50 V and the PV voltage at MPPT is 50 V, Fig. 5.13 shows the experiment results when the output current i_{PV} of PV panel is changed. As shown in Fig. 5.13, when the output current of the PV panel is changed between 1 and 1.6 A, the total dc-link voltage U_{MD} can be remained at its desired value by using the proposed high-robustness control scheme. Thus, when the output current of the PV panel is changed, excellent dynamic performance can be obtained by using the DAB-based PPP converter system with the proposed control scheme.

Moreover, when the load resistor is 40 Ω, the voltage of the LVDC bus is 50 V and the PV current at MPPT is 1.6 A, Fig. 5.14 shows the experiment result when the terminal voltage U_{PV} of the PV panel at MPPT is changed. As shown in Fig. 5.14, when

Table 5.2 Circuit parameters of the small-scale experiment platform

Parameter	Value
L_1, L_2	40 μH
n_1, n_2, n_3	1
f_s	40 kHz
R	40~90 Ω
U_{LD}	50~60 V
U^*_{MD}	100 V

Fig. 5.12 The configuration of the small-scale experiment platform

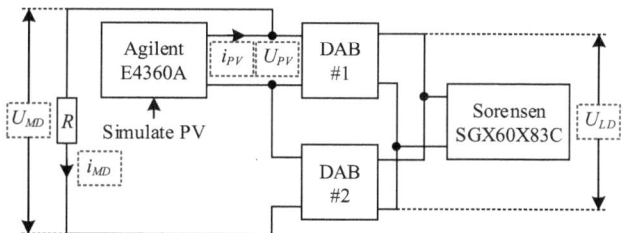

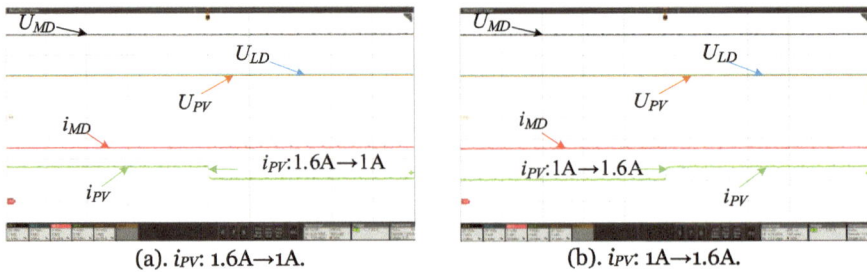

(a). i_{PV}: 1.6A→1A.   (b). i_{PV}: 1A→1.6A.

Fig. 5.13 Experiment results when the output current of the PV panel at MPPT is changed. (U_{MD}, U_{PV} and U_{LD}: 25 V/div; i_{MD} and i_{PV}: 1 A/div; t: 20 ms/div)

the output voltage of the PV panel is changed between 45 and 50 V, the total dc-link voltage U_{MD} can be remained at its desired value by using the proposed high-robustness control scheme. Thus, when the terminal voltage of the PV panel is changed, excellent dynamic performance can be obtained by the proposed control scheme. Combining the experiment results as shown in Figs. 5.13 and 5.14, the variation of the PV irradiance can be simulated, and the excellent dynamic performance can be obtained by using the proposed high-robustness control scheme under this condition.

In addition, when the voltage of the LVDC bus is 50 V, the PV voltage is 50 V and the PV current at MPPT is 1.6 A, Fig. 5.15 shows the experiment result when the load resistor R is changed. As shown in Fig. 5.15, when the load resistor is changed between 40 and 90 Ω, the total dc-link voltage U_{MD} can be remained at its desired value by using the proposed high-robustness control scheme. Thus, when the load resistor is changed, excellent dynamic performance can be obtained by the proposed control scheme for the DAB-based PPP converter system.

Similarly, when the load resistor is 40 Ω, the PV voltage is 50 V and the PV current at MPPT is 1.6 A, Fig. 5.16 shows the experiment result when the terminal voltage U_{LD} of the LVDC bus is changed. As shown in Fig. 5.16, when this voltage is changed between

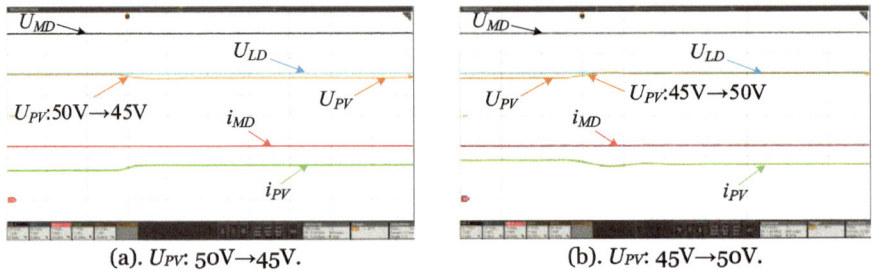

(a). U_{PV}: 50V→45V.   (b). U_{PV}: 45V→50V.

Fig. 5.14 Experiment results when the terminal voltage of PV panel at MPPT is changed. (U_{MD}, U_{PV} and U_{LD}: 25 V/div; i_{MD} and i_{PV}: 1 A/div; t: 100 ms/div)

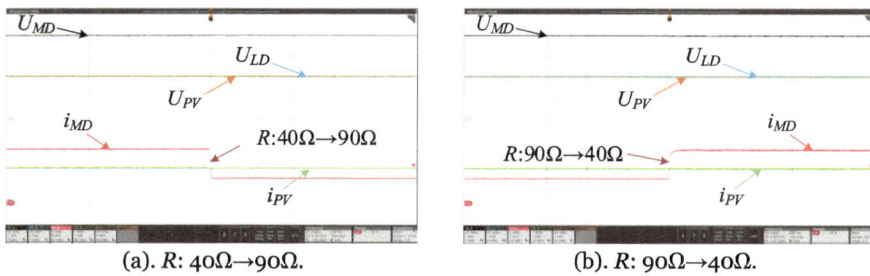

(a). R: 40Ω→90Ω.   (b). R: 90Ω→40Ω.

Fig. 5.15 Experiment results when the load resistor is changed. (U_{MD}, U_{PV} and U_{LD}: 25 V/div; i_{MD} and i_{PV}: 1 A/div; t: 20 ms/div)

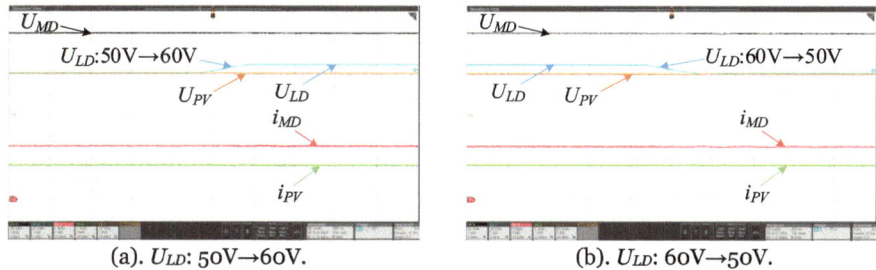

(a). U_{LD}: 50V→60V.   (b). U_{LD}: 60V→50V.

Fig. 5.16 Experiment results when the voltage of the LVDC bus is changed. (U_{MD}, U_{PV} and U_{LD}: 25 V/div; i_{MD} and i_{PV}: 1 A/div; t: 20 ms/div)

50 and 60 V, the total dc-link voltage U_{MD} can be remained at its desired value by using the proposed high-robustness control scheme. Thus, when the terminal voltage of the LVDC bus is changed, excellent dynamic performance can be obtained by the proposed control scheme for the DAB-based PPP converter system.

Furthermore, when the load resistor is 40 Ω, the voltage of the LVDC bus is 50 V, the output current of PV is 1.6 A and the PV voltage at MPPT is 50 V, Fig. 5.17 shows the experiment results to simulate the transient processes when the PV panels are heavily shaded or activated again. As shown in Fig. 5.17a, when the PV panel is heavily covered, the output current is suddenly become zero, and based on the high-robustness control scheme with the discussed operation in Sect. 5.2.1, the total dc-link voltage U_{MD} can be kept at its desired value. Besides, as shown in Fig. 5.17b, when the PV panel is activated again, the total dc-link voltage U_{MD} is also constant. Therefore, based on the proposed high-robustness control scheme with discussed operation in Sect. 5.2.2, when the PV panel is heavily shaded or activated suddenly, the excellent dynamic performance can be provided for the presented DAB-based PPP converter system.

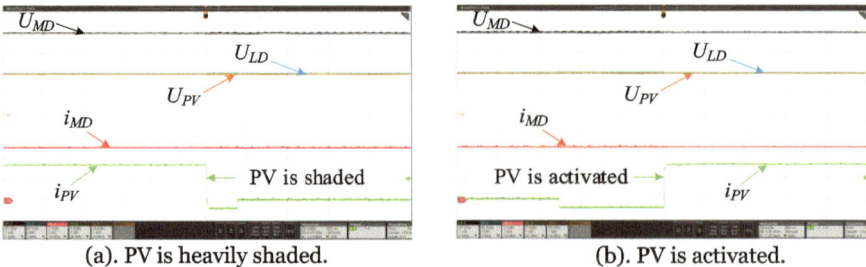

(a). PV is heavily shaded.   (b). PV is activated.

Fig. 5.17 Experiment results when the PV panel is heavily shaded or activated again. (U_{MD}, U_{PV} and U_{LD}: 25 V/div; i_{MD} and i_{PV}: 1 A/div; t: 20 ms/div)

5.4   Summary

In this chapter, a DAB-based PPP converter with robust dc-link voltage is analyzed for the islanded dc microgrid embedded with the renewable energy source and the energy storage system. Based on the proposed PPP converter system, the independent control of the renewable energy source and the stabilization of the total dc-link voltage can be realized. Moreover, for this DAB-based PPP converter system, a high robustness control strategy is proposed for maintaining the dc-link voltage under some cases: (1) The working condition of the renewable energy source is changed. (2) The voltage of the battery is varied. (3) The load condition is changed. In addition, when the renewable energy source is out of work, the robustness of the total dc-link voltage can be ensured by combining a presented operation. Notably, renewable energy should feature the current output and the limited output-voltage regulation such as PV, fuel cell and wind turbine with ac-dc conversion. Furthermore, by using the PV as example, the effectiveness of the proposed DAB-based PPP converter system and the proposed high-robustness control scheme is verified: (1) The MPPT of the PV panels can be realized by the existing method. (2) The total dc-link voltage can maintain at its desired value when the irradiance of PV panels, the voltage of the battery, and the load condition are changed, and even when the PV panel is heavily shaded.

References

1. W. Song, N. Hou, and M. Wu, "Virtual Direct Power Control Scheme of Dual Active Bridge DC–DC Converters for Fast Dynamic Response," *IEEE Transactions on Power Electronics,* vol. 33, no. 2, pp. 1750–1759, 2018.
2. N. Vazquez, and M. Liserre, "Peak Current Control and Feed-Forward Compensation of a DAB Converter," *IEEE Transactions on Industrial Electronics,* vol. 67, no. 10, pp. 8381–8391, 2020.

3. M. G. Villalva, J. R. Gazoli, and E. R. Filho, "Comprehensive Approach to Modeling and Simulation of Photovoltaic Arrays," *IEEE Transactions on Power Electronics,* vol. 24, no. 5, pp. 1198–1208, 2009.

4. H. A. Sher, A. F. Murtaza, A. Noman, K. E. Addoweesh, K. Al-Haddad, and M. Chiaberge, "A New Sensorless Hybrid MPPT Algorithm Based on Fractional Short-Circuit Current Measurement and P&O MPPT," *IEEE Transactions on Sustainable Energy,* vol. 6, no. 4, pp. 1426–1434, 2015.

5. S. Bhattacharyya, D. S. K. P, S. Samanta, and S. Mishra, "Steady Output and Fast Tracking MPPT (SOFT-MPPT) for P&O and InC Algorithms," *IEEE Transactions on Sustainable Energy,* vol. 12, no. 1, pp. 293–302, 2021.

6. M. A. G. d. Brito, L. Galotto, L. P. Sampaio, G. d. A. e. Melo, and C. A. Canesin, "Evaluation of the Main MPPT Techniques for Photovoltaic Applications," *IEEE Transactions on Industrial Electronics,* vol. 60, no. 3, pp. 1156–1167, 2013.

7. S. K. Kollimalla, and M. K. Mishra, "Variable Perturbation Size Adaptive P&O MPPT Algorithm for Sudden Changes in Irradiance," *IEEE Transactions on Sustainable Energy,* vol. 5, no. 3, pp. 718–728, 2014.

Conclusions and Future Plans

6

The main objectives of this thesis can be divided into three parts:

(1) The dynamic equivalence between the DAB dc-dc converter and other I^2ACL dc-dc converters should be verified. Then, this verification can be the basis for extending the dynamic optimization schemes of the modular DAB dc-dc converter systems such as the IPOP, IIOP, IPOS and ISOP configurations to other I^2ACL dc-dc converter with the same configurations. Besides, a unified fast-dynamic control should be proposed for these I^2ACL dc-dc converters for boosting the dynamic response when the input voltage and the load resistor are changed.

(2) Currently, the modular DAB dc-dc converter systems such as the IPOP, IIOP, IPOS and ISOP DAB dc-dc converter systems start to get the attentions in the dc power conversion system for realizing high power transmission and connecting different voltage-level dc terminals. The fast-dynamic response is a crucial requirement when these modular DAB dc-dc converter systems should provide a constant dc voltage for the following consumers. Thus, the fast-dynamic control schemes for these DAB-based converter systems should be studied when the input voltage, the load condition and the power sharing performance are changed. Moreover, the hot-swap operations should be researched for switching the DAB module, and the CPE method should also be discussed for ensuring the desired power sharing performance of these modular DAB dc-dc converter systems.

(3) Although the PPP converter systems are extensively studied for connecting the renewable energy source and the strong ac grid system, the PPP converter system for the islanded dc microgrid still needs study. To expand the scope of the application of the DAB-based converter system, the DAB-based PPP converter system with robust

N. Hou, *High-Robust Control Schemes for Dual-Active-Bridge-Based DC–DC Converter Systems in Renewable Energy Applications*, Synthesis Lectures on Power Electronics, https://doi.org/10.1007/978-3-031-72963-8_6

dc-link voltage should be studied for the islanded dc microgrid with the renewable energy and the energy storage system. Notably, this proposed converter system should realize the independent control of the renewable energy source and the stabilization of the total dc-link voltage. Moreover, the high-robustness control strategy and the source-shaded operation should be researched to boost the robustness of this novel DAB-based PPP converter system.

6.1 Achievements of This Work

In this work, the thesis contributes to the field DAB-based DAB dc-dc converter systems in four areas.

(1) The Unified Fast-Dynamic Control Scheme for I^2ACL Isolated dc-dc Converters.

In Chap. 2, the existing isolated dc-dc converters with I^2ACL feature are reviewed since these existing converters have similar dynamic characteristics. Although there is an ac inductance in the middle of these converters, this work reveals that this ac inductance doesn't influence the power transmission in each switching period. Thus, these I^2ACL dc-dc converters can be treated as the first-order converter. Based on these analyses, the dynamic equivalence between the DAB dc-dc converter and other I^2ACL dc-dc converters can be verified. Moreover, a unified FDDC control scheme is proposed for providing the ultrafast dynamic response for these dc-dc converters when the input voltage and the load condition are changed. In addition, because of the first-order power transferred character, the unique design principle of PI parameters is also presented for these converters. Furthermore, with the dynamic equivalence among these isolated dc-dc converters, the following proposed control concepts for the modular DAB dc-dc converters including the IPOP, IIOP, IPOS, and ISOP DAB dc-dc converter systems can be easily extended to other I^2ACL dc-dc converters with the same configuration.

(2) The Flexible Power Sharing Control Schemes with Fast-Dynamic Response for the IPOP DAB dc-dc Converter System and the IIOP DAB dc-dc Converter System.

In Chap. 3, a tunable power sharing control scheme for the IPOP DAB dc-dc converter is proposed for flexible power management of the DAB modules. Besides, excellent dynamic performance under disturbances of the input voltage and the load resistor can be provided. Moreover, to ensure the desired power sharing performance, the comprehensive CPE schemes proposed for different conditions of the IPOP DAB dc-dc system including the start-up process, the working process and plugging-in a new DAB dc-dc converter, respectively. In addition, a communication-free power management strategy is proposed for the IIOP DAB dc-dc converter. Based on the proposed scheme, the stable dc-link voltage can be obtained when the input voltage of each module, the load condition, and the power sharing performance are changed. Furthermore, the hot swap (plug-in and plug-out) control methods for the IPOP DAB dc-dc converter and the IIOP DAB dc-dc converter without large influence on output voltage are presented in detail.

(3) The Flexible Power Sharing Control Schemes with Fast-Dynamic Response for the IPOS DAB dc-dc Converter System and the ISOP DAB dc-dc Converter System.

In Chap. 4, this work proposes a simple tunable power sharing control strategy for the IPOS DAB dc-dc converter system. Based on the proposed scheme, excellent dynamic performance can be obtained when the input voltage, the load resistor, and the power sharing performance are changed. Besides, based on a variant of the proposed scheme, the black-start operation can be obtained for synchronous charging of the output capacitors during start-up process. Moreover, the hot-swap operation is presented for the IPOS DAB dc-dc converter with a slight influence on the output voltage, and this hot-swap concept can also be used for the ISOP DAB dc-dc converter system. In addition, an input-oriented power sharing control scheme with fast-dynamic response is proposed for the ISOP DAB dc-dc converter system. Based on this proposed scheme, the complete decoupling between the regulations of the input voltages and the output voltage can be realized. Further, excellent dynamic performance can be obtained when the input voltage, the load resistor and the power sharing performance are changed. In addition, an inductance-estimating method is proposed for ensuring the power sharing performance of the ISOP DAB dc-dc converter, which can also be employed in the IPOS DAB dc-dc converter system.

(4) The Partial Power Processing Converter System with Robust DC-Link Voltage for Islanded DC Microgrid

In Chap. 5, the proposed DAB-based PPP converter system is analyzed for the islanded dc microgrid with the renewable energy and the energy storage system. This DAB-based PPP converter system can be employed to realize the independent control of the renewable energy sources and the constant total dc-link voltage simultaneously. Moreover, for this novel DAB-based PPP converter system, a high robustness control strategy is proposed for maintaining the dc-link voltage under some cases including the changes of the renewable energy source, the load resistor, and the battery voltage. In addition, when the renewable energy source is out of work, the stability of the total dc-link voltage can be ensured by combining a presented operation. Notably, renewable energy should feature the current output and the limited output-voltage regulation such as PV, fuel cell and wind turbine with ac-dc conversion. By using the PV as an example, the effectiveness of the proposed DAB-based PPP converter system can be verified. Further, the effectiveness of the proposed high-robustness control scheme is verified since the total dc-link voltage can maintain at its desired value when the irradiance of PV panels, the voltage of the battery and the load condition are changed, and even when the PV panel is heavily shaded.

6.2   Future Work

The future works are suggested to focus on the multiple operations of the proposed PPP converter system with or without stiff ac grid, where the inverter is employed to connect the proposed PPP converter system and the ac grid. Moreover, to complete the dc distributed system, the connection between the dc voltage terminal and the dc current terminal should be studied for interfacing dc current transmission line such as high-voltage direct-current transmission and the dc microgrid. In addition, the input-series output-series DAB dc-dc converter system is also an important application of the DAB-based converter system, so the flexible power sharing control scheme with fast-dynamic control scheme should be studied for this input-series output-series DAB dc-dc converter system.

(1) *The Multiple Operations of the PPP Converter System with or Without Strong Ac Grid*

In this work, the PPP converter system with robust dc-link voltage is proposed for the islanded dc microgrid with the renewable energy source and the energy storage system. However, the strong ac system is usually employed to support the electricity consumer in power system. Thus, based on inverter, the proposed PPP converter system can be connected to the ac grid. Then, the power management of the proposed PPP converter system with or with stiff dc grid should be further studied for better interfacing the renewable energy, the energy storage system, and the strong ac system.

(2) *Hybrid-Type DC-DC Converters Interfacing DC-Current Buses and DC-Voltage Buses*

Along with the application of the current source converter, there are more and more dc-current buses for supporting power conversion, and in the current power system, the dc-voltage bus is the most dc terminal. Thus, it is foreseeable that topologies which can connect the dc-current bus and the dc-voltage bus will be urgently needed in the future grid system, such as the dc grid and electric vehicle. So, the hybrid-type dc-dc converter concept interfacing the dc-current bus and the dc-voltage bus should be studied, and the potential hybrid dc-dc converters without or with electric isolation should be obtained. Moreover, for these hybrid dc-dc converters, the modulation operations should be proposed and analyzed.

(3) *The Flexible Power Sharing Control Scheme for the Input-Series Output-Series DAB DC-DC Converter System*

In this work, the flexible power sharing control schemes with fast-dynamic response for most of the modular DAB dc-dc converter systems including the IPOP, IIOP, IPOS and ISOP DAB dc-dc converter systems are proposed for achieving the excellent dynamic performances when the input voltage, the load condition and the power sharing performance are changed. However, the flexible power sharing control scheme is not proposed for the input-series output-series DAB dc-dc converter system. The input-series output-series DAB dc-dc converter system can be employed to connect two high-voltage dc terminals, and when this modular DAB dc-dc converter is employed to provide power to the consumer side, the robustness of the output voltage is also a crucial requirement. Thus, a flexible power sharing control schemes with fast-dynamic response should be provided for the input-series output-series DAB dc-dc converter system when the input voltage, the load condition and the power sharing performance are changed.

Index

B

Black-start control, 17, 103, 121, 128, 149

C

Control extension, 77

D

DAB-based PPP converter system, 2, 14–17, 131, 132, 134–139, 141–144, 147–149

DAB converter, 6, 64, 65, 68, 70–72, 75, 78, 81, 86, 91, 93, 95, 99, 100, 105, 106, 113, 115, 131, 133, 134

Dynamic equivalence, 1, 2, 4, 5, 16, 58, 147, 148

Dynamic response, 10, 12, 15–17, 41, 45, 79, 80, 86, 103, 110, 112, 115–117, 120, 124, 126–128, 131, 137, 141, 147–149, 151

E

Energy storage system, 1, 7, 8, 11, 14, 40, 100, 144, 148–150

H

High-robust control, 14, 17, 131, 134–136, 141–144, 148, 149

Hot-swap operation, 5, 8, 16, 63, 64, 77, 100, 128, 147, 149

I

I^2ACL isolated DC-DC converter, 2, 5, 16, 27–29, 40, 43, 48, 148

Input-Independent Output-Parallel (IIOP), 1, 2, 5, 8, 9, 16, 17, 58, 63, 77, 79, 80, 85, 93, 95, 97, 98, 100, 147–149, 151

Input-Parallel Output-Parallel (IPOP), 1, 2, 5, 6, 12, 16, 58, 63–69, 71–73, 85, 86, 88, 89, 91, 92, 100, 147–149, 151

Input-Parallel Output-Series (IPOS), 1, 2, 9, 10, 16, 17, 58, 103–105, 107, 108, 120, 122, 128, 147–149, 151

Input-Series Output-Parallel (ISOP), 1, 2, 9, 11, 12, 14, 16, 17, 58, 103, 110, 111, 115–120, 124, 127–129, 147–149, 151

Islanded DC microgrid, 2, 7, 8, 14, 15, 17, 144, 147, 149, 150

Intermediary inductive AC-link converters, 1

P

Power sharing control, 5, 6, 8, 11, 12, 16, 17, 63, 64, 69, 73, 74, 86, 88, 91, 92, 100, 103, 107–110, 112, 113, 115–117, 120, 122, 124, 127, 128, 148–151

R

Renewable energy, 2, 7, 13, 14, 16, 17, 131, 134, 136, 137, 144, 147–150

N. Hou, *High-Robust Control Schemes for Dual-Active-Bridge-Based DC–DC Converter Systems in Renewable Energy Applications*, Synthesis Lectures on Power Electronics, https://doi.org/10.1007/978-3-031-72963-8